Embracing Microservices Design

A practical guide to revealing anti-patterns and architectural pitfalls to avoid microservices fallacies

Ovais Mehboob Ahmed Khan

Nabil Siddiqui

Timothy Oleson

BIRMINGHAM—MUMBAI

Embracing Microservices Design

Group Product Manager: Richa Tripathi
Publishing Product Manager: Alok Dhuri
Senior Editor: Rohit Singh
Content Development Editor: Kinnari Chohan
Technical Editor: Karan Solanki
Copy Editor: Safis Editing
Project Coordinator: Deeksha Thakkar
Proofreader: Safis Editing
Indexer: Tejal Daruwale Soni
Production Designer: Jyoti Chauhan

First published: September 2021

Production reference: 2221021

Published by Packt Publishing Ltd.
Livery Place
35 Livery Street
Birmingham
B3 2PB, UK.

ISBN 978-1-80181-838-4

www.packt.com

Foreword

In the last 5 years, the microservice architecture design approach has established itself in the software engineering industry as the primary way to build modern, scalable, resilient services. This has been driven by the wave of cloud adoption and a major shift toward "cloud-native" service development, often built with microservice architectures (both stateless and stateful) that run on the cloud and the edge and embracing the diversity of programming languages and frameworks.

It's not that other architectures' design approaches, such as client/server, are bad; they just address a different need focusing on medium- to large-scale organizations for their scale and resilience. In the far more competitive services world that has emerged, where delivering new features on a regular basis is expected, and a continuous integration and delivery approach is required, the microservices architecture has become the de facto standard for building them. However, a microservice architecture alone was not enough to drive this adoption without suitable distributed systems hosting platforms to run these services on that provided the horizontal scaling, automated rollouts, load balancing, and self-healing, such as Kubernetes, Docker Swarm, and Azure Service Fabric. These infrastructure platforms and the newer serverless container platforms have been instrumental in the rise of the microservice architecture approach.

This book takes you along the journey of understanding, using, and applying microservice design principles and patterns, consolidated into a comprehensive set of guidance and supported by the authors' real-world experiences. Chapters such as *Chapter 10, Evaluating a Microservice Architecture,* enable you to see the bigger picture of your goals, with the other chapters diving into the best approach and pitfalls to watch out for. Although microservices may seem daunting at first, with the need for service discovery, observability, security, and separated state deployed across machines in a distributed manner, there are tools, frameworks, and runtimes that have emerged to greatly simplify the burden on developers. These include many of the Cloud Native Computing Foundation (cncf.io) technologies, and the Distributed Application Runtime (dapr.io), covered in *Chapter 3, Microservices Architecture Pitfalls*, which greatly ease the burden for developers, by codifying best practices into APIs for use from any language.

I am confident that you will find this book invaluable in your microservices design journey and use it frequently to design applications for the modern services era of computing.

Mark Fussell
Microsoft Partner Program Manager
Founder of the **Distributed Application Runtime (Dapr)** and Azure Service Fabric

Contributors

About the authors

Ovais Mehboob Ahmed Khan is a seasoned programmer and solution architect with nearly 20 years of experience in software development, consultancy, and solution architecture. He has worked with various clients across the world and is currently working as a senior customer engineer at Microsoft. He specializes mainly in application development using .NET and OSS technologies, Microsoft Azure, and DevOps. He is a prolific writer who has written several books and articles on various technologies. He really enjoys talking about technology and has given a number of technical sessions around the world.

Nabil Siddiqui is an open source contributor and technology speaker who is currently working as a cloud solution architect at Microsoft, focusing on application innovation. He is well versed as a software engineer, consultant, trainer, and architect. He has around 20 years of experience, ranging from desktop applications to highly distributed applications on different platforms using various programming languages. During his career, he's accomplished various technical challenges and worked on a broad range of projects that have given him rich experience in designing innovative solutions to solve difficult problems.

Tim Oleson, a customer engineer at Microsoft, focuses on domain-driven design, Cosmos DB, Microsoft Azure PaaS, and Azure Kubernetes Service. He is a seasoned software engineer, consultant, trainer, and architect. He has 20+ years of experience, ranging from web applications to highly distributed service-oriented applications.

About the reviewers

Ovais Khan is a seasoned software engineer and technical leader who has architected, built, and operated a number of microservices. Ovais has extensive experience of building and operating large-scale distributed systems, cloud-scale database systems, and multi-cloud systems. He is currently working as a software engineer at Snap Inc., where he has worked on NoSQL storage solutions and live data migrations and transitioned a number of features from a monolith to microservices. Prior to that, Ovais worked at Microsoft, where he was part of the Azure Storage, Office 365, and SharePoint teams.

David McGhee is a technical program manager at Microsoft in the Commercial Software Engineering team focusing on intelligence within government in Australia. He has a sustained performance over 28 years in many distributed applications, from broadcast television to real-time financial services. David has led many teams and embraced technology firsts, including Azure in its infancy. This background has inevitably led to many failures and learnings, even successes, in cloud solution architecture. He enjoys the charm of conversation this discourse creates and opportunities for teams to get better.

Table of Contents

2
Failing to Understand the Role of Domain-Driven Design

3
Microservice Architecture Pitfalls

4
Keeping Replatforming Brownfield Applications Trivial

Section 2: Overview of Data Design Pitfalls, Communication, and Cross-Cutting Concerns

5

Data Design Pitfalls

6

Communication Pitfalls and Prevention

7

Cross-Cutting Concerns

Section 3: Testing Pitfalls and Evaluating Microservices Architecture

8

Deployment Pitfalls

9

Skipping Testing

10

Evaluating a Microservices Architecture

Assessments

Other Books You May Enjoy

Index

Preface

Embracing Microservices Design targets architects and developers who want to design and build microservices-based applications following the right principles, practices, and patterns and avoiding the fallacies and anti-patterns that may lead to bad implementation. The adoption of microservices is growing due to various benefits, such as agility, maintainability, technology diversity, scalability, innovation, faster time to market, and increased reliability. However, it comes with many challenges, such as a change in the organization's culture, a learning curve, understanding and modeling business domains, architecture and operational complexity, observability, and end-to-end testing.

This book is divided into three sections. The first section is focused on a microservices introduction and looking at design and architecture pitfalls followed by an alternative approach to avoid those pitfalls. The second section is focused on data design pitfalls, communication challenges and preventions, and cross-cutting concerns. The last section is focused on deployment pitfalls and best practices when setting up the deployment strategy, the need for testing, challenges, and how to evaluate the potential application when transforming from a monolithic architecture to a microservices architecture.

Who this book is for

This book is for architects and developers who are involved in designing microservices and related architectures. Working proficiency with microservices is assumed to get the most out of this book.

What this book covers

Chapter 1, Setting Up Your Mindset for Microservices Endeavor, teaches you about the role of different individuals while initiating a microservices endeavor, understanding the importance of building teams, and investing in your learning to execute the microservices adoption strategy according to the defined charter. This chapter also discusses the guidelines for building scalable, maintainable, and portable applications while adopting the 12-Factor App methodology.

Chapter 2, Failing to Understand the Role of DDD, teaches you about the importance of understanding domain-driven design to build cohesive services with bounded context. This chapter will re-examine the characteristics and properties of microservices with respect to domain-driven design and how it relates to building microservices. You will also learn about the importance of team building, governance, and the awareness of stakeholders.

Chapter 3, Microservices Architecture Pitfalls, teaches you about various architecture pitfalls while building microservices. It covers the complexities and benefits of microservices by relating them to various architecture practices and how they can be improved by using different patterns. Furthermore, it discusses the importance of understanding the pros and cons of various frameworks and technologies. Finally, it emphasizes the benefits of abstracting common microservices tasks and the drawback of a monolithic frontend.

Chapter 4, Keeping the Replatforming of Brownfield Applications Trivial, teaches you the techniques of replatforming brownfield applications followed by some patterns that can be used. Transitioning from a monolithic to a microservices architecture is not a straightforward process and there are many factors to consider, such as availability, reliability, and scalability, which are also discussed later in this chapter. Finally, it discusses some of the drawbacks of reusing monolithic application components rather than rewriting them to take advantage of emerging technologies that may offer better alternatives.

Chapter 5, Data Design Pitfalls, teaches you the importance of decomposing the database of a monolithic application into a set of microservices where each microservice manages its own database. Normalizing data in microservices could be an anti-pattern in many scenarios and this chapter discusses the reasons this should be avoided. This chapter covers strategies to break data out of monolithic databases and examines the pitfalls of not knowing how to handle transactions. You will also learn about atomic transactions across a distributed architecture using sagas. Finally, this chapter covers the pitfalls of not knowing how to perform reporting by implementing an API that is responsible for building complex reports for business analysis.

Chapter 6, Communication Pitfalls and Prevention, teaches you about various communication styles for microservices architecture, their challenges, and potential solutions. You will also learn about different techniques to enable resilient inter-process communication between microservices. Furthermore, it discusses the challenges of event-driven architecture and how they can be addressed using different approaches. Finally, you will learn about service meshes and how they compare to Dapr.

Chapter 7, Cross-Cutting Concerns, teaches you about the importance of the microservices chassis pattern and its use cases. This chapter starts with discussing some common needs for every microservices-based application and the pitfalls when addressing these concerns.

Chapter 8, Deployment Pitfalls, teaches you about various deployment pitfalls by starting with the necessity of having different deployment techniques. This chapter discusses the importance of using the right and dated tools and technologies to avoid failures. After that, you will learn about DevOps practices and factors to be considered, such as feature management, agile practices, a rollback strategy, approvals, and gates, when implementing it. Finally, this chapter covers various deployment patterns, such as deployment stamps, deployment rings, and geodes, and where they can be applied.

Chapter 9, Skipping Testing, teaches you about the importance of testing to the development and deployment of microservices. Skipping testing is not a good practice and can lead to failure. However, this can be avoided by shifting your testing to the left of your development life cycle. This chapter covers some strategies related to shift-left testing and the types of testing that support this approach.

Chapter 10, Evaluating Microservices Architecture, teaches you about some essential factors to consider while evaluating a microservices architecture. Giving these factors high priority while developing or replatforming an app will help you identify areas where improvement needs to happen.

Download the color images

We also provide a PDF file that has color images of the screenshots and diagrams used in this book. You can download it here: `https://static.packt-cdn.com/downloads/9781801818384_ColorImages.pdf`.

Conventions used

There are a number of text conventions used throughout this book.

`Code in text`: Indicates code words in text, database table names, folder names, filenames, file extensions, pathnames, dummy URLs, user input, and Twitter handles. Here is an example: "The invoice generation can subscribe to `payment_recieved` and perform the necessary action to generate the invoice. Once the invoice is generated, the system will generate an `invoice_generated` event, which can be processed by the notification system to send notifications."

Bold: Indicates a new term, an important word, or words that you see onscreen. For instance, words in menus or dialog boxes appear in **bold**. Here is an example: "**Trusted Host** is the system or platform used to host the microservices."

> **Tips or important notes**
> Appear like this.

Get in touch

Feedback from our readers is always welcome.

General feedback: If you have questions about any aspect of this book, email us at customercare@packtpub.com and mention the book title in the subject of your message.

Errata: Although we have taken every care to ensure the accuracy of our content, mistakes do happen. If you have found a mistake in this book, we would be grateful if you would report this to us. Please visit www.packtpub.com/support/errata and fill in the form.

Piracy: If you come across any illegal copies of our works in any form on the internet, we would be grateful if you would provide us with the location address or website name. Please contact us at copyright@packt.com with a link to the material.

If you are interested in becoming an author: If there is a topic that you have expertise in and you are interested in either writing or contributing to a book, please visit authors.packtpub.com.

Share Your Thoughts

Once you've read *Embracing Microservices Design*, we'd love to hear your thoughts! Scan the QR code below to go straight to the Amazon review page for this book and share your feedback.

https://packt.link/r/1-801-81838-X

Your review is important to us and the tech community and will help us make sure we're delivering excellent quality content.

Section 1: Overview of Microservices, Design, and Architecture Pitfalls

In this section, you'll learn about the fundamentals of microservices, the importance of domain-driven design, architecture pitfalls, and the decomposition strategy. This section comprises the following chapters:

- *Chapter 1, Setting Up Your Mindset for Microservices Endeavor*
- *Chapter 2, Failing to Understand the Role of DDD*
- *Chapter 3, Microservices Architecture Pitfalls*
- *Chapter 4, Keeping the Replatforming of Brownfield Applications Trivial*

1
Setting Up Your Mindset for a Microservices Endeavor

Microservices is an architectural style that structures an application into multiple services that encapsulate business capabilities. These microservices are usually much smaller, both in terms of scope and functionality, compared to traditional services. They have logical and physical separation, and they communicate with each other via messages over a network to form an application.

Adopting the microservice architecture is a journey that requires a mindset that can embrace changes in culture, processes, and practices. In essence, organizations need to go through several changes to increase agility and adaptability to achieve their goals and deliver business value.

In this chapter, we will help you understand the philosophy of microservices and the various aspects that help organizations with their digital transformation. We will also learn about microservice design principles and their components, along with their benefits and challenges. Finally, we will discuss the role of leadership and how to get started by adopting widely accepted practices in the industry.

The following topics will be covered in this chapter:

- Philosophy of microservices
- Microservice design principles
- Building teams to deliver business value faster
- Benefits of microservices
- Challenges of microservices
- Microservice architecture components
- Reviewing leadership responsibilities
- Defining core priorities for a business
- Using the twelve-factor app methodology
- Additional factors for modern cloud-native apps

This chapter aims to help you understand the key concepts of microservices and their potential for bringing change that can fuel business growth for organizations. In addition, we will discuss the role of leadership in shifting the mindset by building effective teams that embrace change to deliver business value faster. Finally, we will dive into the "Twelve-Factor App" methodology, which has helped many organizations in their successful adoption of microservices.

Philosophy of microservices

The digital revolution is disrupting every industry to fulfill the unmet demands of its users and embrace digital transformation to transform their current businesses. Digital transformation allows businesses to adapt quickly to the changing business conditions and create value for their end users. Digital transformation is about empowering your employees, engaging your customers, transforming products, and optimizing operations. The essence of digital transformation is in a growth mindset, where organizations invest in improving their internal capabilities, processes, and systems to bring change that drives business and societal outcomes. Many of these systems are due for a change that would enable them to innovate at a rapid pace and reimagine the future to deliver exciting experiences. In the last few decades, most of these systems were built using a monolithic architecture, which is relatively difficult to change and maintain as it grows and becomes complex. Though the Service-oriented architecture (SOA) was a major improvement over the monolithic architecture, it has its own challenges, such as when the application grows, it becomes a distributed monolithic application, which is again difficult to maintain or extend. The focus of SOA is more on reusability and integration, where services constantly exchange messages in return for tasks to be executed. The messaging platform is the backbone of SOA and is responsible for service discovery, orchestration, routing, message transformation, message enrichment, security, and transaction management. The major downside of SOA is that the information is shared and knowledge of the business domain is scattered across the **Enterprise Service Bus** (**ESB**), which makes it difficult to change the service.

Microservices is an evolution of the architecture style that addresses the pain points of other architecture styles to enable rapid change and scale. Microservices also enable continuous delivery of business capabilities. The effort of concentrating on and segregating business capabilities from each other in individual services enables organizations to build systems that are modular, isolated, and loosely coupled in nature. These characteristics play a crucial role in helping organizations build dedicated teams that are focused on delivering engineering velocity. Teams are isolated to develop and deploy microservices independently, without any major collaboration required. However, if there are dependencies between services and the services haven't been modeled properly, they undergo extensive collaboration, violating the isolation as a benefit. The microservice architecture also enables organizations to build services with autonomy that embraces change and lowers the risk of failures because of service independence.

Microservices have changed the application development paradigm, where new fundamentals have surfaced and are successful at building and operationalizing microservices. Organizations need to start with the right mindset and build a culture of ownership across small teams to continuously deliver value to their end users.

Now that you have a basic understanding of the philosophy behind microservices, next, we will go through microservice design principles.

Microservice design principles

In the quest of building microservices, you will have to make several choices that will give a different flavor to your microservice architecture and its evolution over time. Microservice design principles provide the guidelines for evaluating key decisions that can affect the design of your microservice-based architecture. In principle, microservices support loose coupling and the modularity of services. Other principles can govern the design of a microservice architecture, but their importance can vary. We will cover those principles in the following sections.

Single responsibility principle and domain-driven design

A microservice should be responsible for offering a single feature or a group of related features to deliver a business capability. The only reason why a microservice interface should change is due to changes in the business capabilities offered by the microservice. This helps systems to be designed following the real-world domains and helps us visualize systems and architectures as a translation of real-world problems. Domain-driven design is an approach that helps with domain modeling and defining microservice boundaries, which helps us achieve modularity and reduces coupling.

Encapsulation and interface segregation

Each microservice owns its data, and the only way a service can communicate with other services is through well-defined interfaces. These interfaces are carefully designed by keeping their clients in mind. Rather than overloading microservice endpoints in client applications, a popular alternative is to introduce an API gateway, which will be explained later in the chapter. This technique is useful in delegating the communication responsibility to API gateways and keeping microservices focused on delivering business capabilities.

Culture of autonomy, ownership, and shared governance

The microservice architecture allows business capabilities owned by different teams to be delivered. These teams can work independently without requiring much collaboration across teams. Each team doesn't need to be assigned a single business capability; instead, they may choose from a related set of capabilities that belong to a single domain. The microservice architecture flourishes when you allow teams to have autonomy, since they can choose what they think is right to deliver the desired business capability. This doesn't mean that teams can do anything, though, but it certainly gives them the freedom to make decisions under an umbrella of formally agreed principles. These principles are called shared governance, which provides consensus across the teams regarding how they want to address different cross-cutting concerns, or how far they want to go to try different technologies. The team that builds the microservice owns it and is responsible for operationalizing it. Such cross-cutting concerns will be covered in detail in *Chapter 7, Cross-Cutting Concerns*.

Independently deployable

A single microservice should be independently deployable, allowing teams to roll out changes without them affecting other microservices. As part of shared governance, teams should continue to look for technologies and practices that can help them achieve better independence. This characteristic is extremely useful for operationalizing microservices at a large scale.

Culture of automation

Automation is an important concept that promotes the idea of finding opportunities for replacing manual steps with scripted programs to achieve consistency and reduce overhead. Continuous integration, continuous deployment, automated testing, and infrastructure automation are all different forms of automation that can help you reduce the overall overhead of managing microservices.

Designing for failures

Microservices are designed to be highly available and scalable. These services do fail but they should be designed to recover fast. The idea is that the failure of one microservice should not affect other microservices, which helps avoid cascading failures and allows the service to recover as fast as possible to restore the required functionality. Designing services for failure requires having a good understanding of user behavior and expectations. The goal is to keep the system responsive in case of service unavailability, alongside reduced functionality. However, service failures are inevitable and require a pragmatic approach that enables experimentation to find system vulnerabilities. This can be achieved with controlled experimentation that creates chaos to determine how the system would behave differently in different circumstances. This approach is classified as chaos engineering.

With chaos engineering, you experiment with specific areas of the system to identify their weaknesses. Chaos engineering is a practice where you intentionally introduce failures to find undiscovered system issues. This way, you can fix them before they occur unexpectedly and affect your business and users. This exercise also helps in understanding the risks and impacts of turbulent conditions, incident response, gaps in observability, as well as the team's ability to respond to incidents. Many tools can be used to perform chaos engineering, such as litmus, Gremlin, and a few others.

Observability

Observability is a capability that you build as part of a microservice architecture to allow you to identify and reason about the internal state of your system. It also helps with monitoring, debugging, diagnosing, and troubleshooting microservices and their core components in a production environment. In a microservice architecture, you collect logs, metrics, and traces to help teams analyze service behavior. Another important aspect of observability is distributed tracing, which helps in understanding the flow of events across different microservices. In practice, metrics are more important than other forms of monitoring. Many organizations invest a sizable number of resources in metrics, ensuring that they are available in real time, and then use them as the main tool for troubleshooting production issues.

In the next section, we will uncover some core fundamentals of microservices to build teams that are independent and autonomous for bringing agility.

Building teams to deliver business value faster

The microservice architecture encourages the idea of building small, focused, and cross-functional teams to deliver business capabilities. This requires a shift in mindset, led by team restructuring. In 1967, Melvin E. Conway introduced the idea:

> *"Any organization that designs a system will inevitably produce a design whose structure is a copy of the organization's communication structure."*
>
> *- Conway's Law*

Therefore, if an organization needs to build software systems that are loosely coupled, then they need to make sure that they build teams that have a minimum dependency to allow them to function with autonomy.

A team is defined as a cohesive and long-lived group of individuals working together on the same set of business problems, owning a problem space/business area, and having sufficient capacity to build and operate. In essence, the team needs to fully understand the piece of software they are building and operating. This helps them build confidence in the software, optimize lead time, and deploy more frequently to bring agility. In practice, the team's size varies, depending on the size and complexity of the service. Adding a new member to a team adds new connections, resulting in communication overhead, which affects productivity. Organizing teams based on features and components is considered a standard approach to building teams to facilitate collaboration and bring cross-functional skills together. Recently, Matthew Skelton and Manuel Pais introduced a structure that lowers the cognitive load of agile teams by presenting fundamental team topologies in their book *Team Topologies*. There are four fundamental team topologies, as discussed here:

- **Stream-aligned team**: The stream-aligned team is a cross-functional team that's aligned to the value stream of the business and is responsible for delivering customer value with minimal dependencies. They know their customers and apply design thinking to understand business requirements to address customer needs. They support their service in production and are focused on continuous improvement and quality of service. The stream-aligned team is long-lived and has complete ownership of the service. They are responsible for building, operating, diagnosing, and supporting these services during their life cycles.

- **Enabling team**: The role of the enabling team is to provide support and guidance to stream-aligned teams in adopting tools, technologies, and practices that can help them perform better. Enabling teams are formed to identify areas of improvement and promote the learning mindset by keeping the wider organization aware of new technologies and best practices. Depending on their purpose, enabling teams can be short-lived or long-lived in an organization. However, their engagement with stream-aligned teams is only for a short period. Once the required capability is achieved, the enabling teams move to support other teams.

- **Complicated subsystem team**: The complicated subsystem teams are formed to address the requirements of complex subsystems, which can add significant cognitive load to stream-aligned teams and affect their ability to stay focused on the business domain. They are responsible for developing and operationalizing these systems to ensure their service availability when it comes to consuming other value streams.

- **Platform team**: The platform team is responsible for providing infrastructure support to enable stream-aligned teams to deliver a continuous stream of value. They ensure that the capabilities offered by the platform are compelling and provide a good developer and operator experience for consuming and diagnosing these platform services. These services should be accessible via APIs, thus enabling self-service and automation. The platform team is aligned to the product life cycle and guarantees that the stream-aligned teams are getting the right support on time to perform their duties. These services are usually bigger and supported by larger teams.

When you adopt a microservice with DevOps, individuals are responsible for handling both the development and operations of a microservice. These microservices are built with the guidance provided by product managers and product owners. In the real world, teams start small until they grow big and add complexity, before being subdivided into multiple teams to ensure autonomy and independence. For example, your application may need to build multiple channels to interact with its customers and want to work with teams with capabilities around iOS and Android, or they may have to add a notification functionality that needs a separate team altogether. In these teams, individuals have a shared responsibility, where design decisions are usually led by senior engineers and discussed with the team before implementation.

Now, let's learn about the advantages that microservices bring.

Benefits of microservices

One of the key benefits of microservices is the "shift of complexity." We managed to shift the complexity of monolithic applications to multiple microservices and further reduce the individual complexity of the microservice with bounded contexts, but we also increased more complexity to operationalize it. This shift in complexity is not bad, as it allows us to evolve and standardize automation practices to manage our distributed systems well. The other benefits of microservices will be discussed next.

Agility

The microservice architecture promotes experimentation, which allows teams to deliver value faster and help organizations create a competitive edge. Teams iterate over a small piece of functionality to see how it affects the business outcome. Teams can then decide if they want to continue with the changes or whether they should discard the idea to try new ones. Teams can develop, deploy, maintain, and retire microservices independently, which helps them become more productive and agile. Microservices give teams the autonomy of trying out new things in isolation.

Maintainability

The microservice architecture promotes building independent, fine-grained, and self-contained services. This helps developers build simple and more maintainable services that are easy to understand and maintain throughout their life cycle. Every microservice has its own data and code base. This helps in minimizing dependencies and increasing maintainability.

Scalability

The microservice architecture allows teams to independently scale microservices based on demand and forecast, without affecting performance and adding significant cost compared to monolithic applications. The microservice architecture also offers a greater deal of parallelism to help with consistent throughput to address increasing load.

Time to market

The microservice architecture is pluggable, so it supports being able to replace microservices and their components. This helps teams focus on building new microservices to add or replace business capabilities. Teams no longer wait for changes from different teams to be incorporated before releasing new business capabilities. Most of the cross-cutting concerns are handled separately, which helps teams in achieving faster time to market. These cross-cutting concerns will be covered in detail in *Chapter 7, Cross-Cutting Concerns*.

Technology diversity

The microservice architecture allows teams to select the right tools and technologies to build microservices rather than locking themselves in with decisions they've made in the past. Technology diversity also helps teams with experimentation and innovation.

Increased reliability

The microservice architecture is distributed in nature, where individual microservices can be deployed multiple times across the infrastructure to build redundancy. There are two important characteristics of the microservice architecture that contribute to the increased reliability of the overall system:

- Isolating failures by running each microservice in its own boundaries
- Tolerating failures by designing microservices to gracefully address the failures of other microservices

So far, we've learned about the benefits of microservices. However, implementing microservices does bring a few challenges that need to be considered. We will look at these in the next section.

Challenges of microservices

In this section, we will discuss the different challenges you may face as you embark on your microservices journey.

Organizational culture

One of the major hurdles in moving to microservices is the organizational culture, where teams were initially built around technical capabilities rather than delivering business capabilities. This requires an evolving organization, restructuring teams, and changing legacy practices.

Adoption of DevOps practices

DevOps provides a set of practices that combines development and operations teams to deliver value. The basic theme is to shorten the development life cycle and provide continuous delivery with high software quality. Adopting DevOps practices is important for any organization to bring changes faster to market, increase deployment frequency, lower the change failure rate, bring a faster mean time to recover, and a faster lead time for change that delivers value to end users. There are various products that support implementing DevOps practices. Azure DevOps is one of the most popular tools that provides an end-to-end DevOps toolchain. Also, ensure that the necessary changes are made to change management processes, including change control and approval processes so that they align with DevOps practices.

Architectural and operational complexity

The microservice architecture is distributed in nature, which presents several challenges compared to a monolithic architecture. There are more moving parts, and more expertise is required for teams to manage them. A few of those challenges are as follows:

- **Unreliable communication across service boundaries**: Microservices are heavily dependent on the underlying network. Therefore, the network infrastructure has to be designed properly and governed to address the needs of communication, as well as to protect the infrastructure from unwelcome events.

- **Network congestion and latency**: Network congestion is a temporary state of a network that doesn't have enough bandwidth to allow traffic flows. Due to this network congestion, different workloads may experience delayed responses or partial failure, resulting in high latency.

- **Data integrity**: In a microservice architecture, a single business transaction may span multiple microservices. Due to any transient or network failure, if any service fails, it may affect some part of the transaction that creates data inconsistencies across different microservices, resulting in data integrity issues.

Service orchestration and choreography

There is no single way of specifying how different services communicate with each other and how the overall system works. Orchestration introduces a single point of failure by controlling the interaction of different services, while choreography promotes the idea of smart endpoints and dump pipes, with a potential downside of introducing cycling dependencies. In choreography, microservices publish and consume messages from the message broker, which helps the overall architecture to be more scalable and fault-tolerant.

Observability

Managing, monitoring, and controlling microservices at scale is a difficult problem. You need to have good observability in place to understand the interaction and behavior of different parts of the system. You need to collect metrics, logs, call stacks, raise alerts, and implement distributed tracing to help you reason about the system.

End-to-end testing

End-to-end testing the microservices is more challenging as it requires reliable and effective communication to bring all the teams on board. End-to-end testing can also hamper your release frequency. Setting up a test environment is difficult and requires coordination across teams.

Double mortgage period

If you are migrating from a monolithic application to a microservice, then you need to live in a hybrid world for a while to support both the legacy and the new application. This is not easy and requires careful planning when it comes to fixing bugs and adding new features, thus affecting agility and productivity.

Platform investment

Organizations need to invest in platform teams that are responsible for building platform services and addressing cross-cutting concerns. These cross-cutting concerns will be covered in detail in *Chapter 7, Cross-Cutting Concerns*. These teams also build tools and frameworks to help other teams get the job done.

The previous sections have given us a good overview of microservices, along with their design principles, advantages, and challenges. We'll now have a look at the various architecture components of microservices and their interaction when it comes to building scalable and robust applications.

Microservice architecture components

Microservices are a loosely coupled set of services that cooperate to achieve a common goal. Besides microservices, there are other components that play a vital role in a microservice architecture. The set of components that help establish the foundation of microservices are shown in the following diagram:

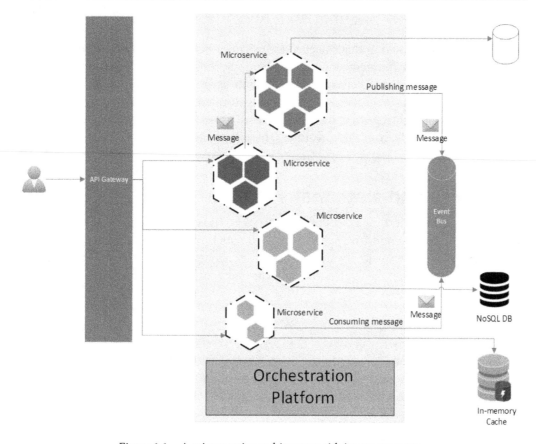

Figure 1.1 – A microservice architecture with its components

The preceding diagram demonstrates the interaction of different microservice architecture components. These microservices are hosted on an orchestration platform, responsible for ensuring self-healing and the high availability of microservices. These microservices then communicate with each other using orchestration or choreography patterns. In orchestration, a microservice is responsible for invoking other microservice interfaces while in choreography, messages are exchanged using an event bus. A client can consume microservices via an API gateway, where messages are relayed to a specific microservice for actual processing. Each team has full autonomy in choosing the right storage mechanism for their microservices, as depicted in the preceding diagram, where different microservices are using a SQL database, an in-memory database, and a NoSQL database for their data storage needs.

We will discuss each component in the following sections.

Messages

Messages contain information that's needed for microservices to communicate with each other. These messages help microservice architecture be loosely coupled. Messages are further classified as commands or events. Commands usually contain more information for the recipient, with the expectation of then being notified about the delivery of messages. Events are lightweight and are mainly used as notification mechanisms, without any expectations from the consumer. These messages can be synchronous or asynchronous.

Persistence and state management

Data handling is an important aspect of microservices. Most of the microservices need to persist data/state. In a microservice architecture, data is decentralized, where every microservice has the responsibility and autonomy of managing its own data. Teams have full autonomy in selecting the right database technology for their microservices to achieve the desired business outcome.

Orchestration

An orchestrator is responsible for placing microservice instances on compute infrastructure. The orchestrator is also responsible for identifying failures and scaling microservices to maintain the high availability and resiliency of the overall architecture.

Service discovery

Service discovery is an integral part of the microservice architecture and helps in locating other microservices to enable collaboration. At a given instance, a microservice may have multiple instances running in a production environment. These instances are dynamically provisioned to address updates, failures, and scaling demands. The service discovery component plays an important role in keeping the microservice architecture highly discoverable by allowing new instances to be registered and become available for service.

API gateway

The API gateway acts as a reverse proxy, responsible for routing requests to the appropriate backend services to expose them in a controlled manner to the outside world for consumption. It also provides robust management and security features that are useful in protecting backend services from malicious actors. The API gateway will be covered in detail in *Chapter 7, Cross-Cutting Concerns*.

Now, let's explore the role of leadership while initiating a microservices endeavor.

Reviewing leadership responsibilities

In this section, we will understand the responsibilities associated with different roles in an organization looking to adopt the culture of continuous delivery of value.

What business/technology leaders must know

Any organization's business leaders must define and communicate the vision of the organization and focus on the necessary grassroot transformation. They must identify growth opportunities, market trends, competition, and address potential risks to create an environment that fuels innovation by bringing new business models to life. Innovation that's driven by a feedback loop creates a culture where new ideas are heard, designed, iterated, and refined. They should focus on value creation through active learning, gaining insights, and experimentation. As business values change over time, this demands a change in business priorities, which should be clearly defined and communicated across the organization so they can be used as the guiding principle to make future decisions.

Technology leaders are responsible for articulating the technology strategy and driving technology initiatives across the organization. They are also responsible for building capabilities by investing in people, processes, products, services, platforms, and acquiring business acumen to deliver a competitive edge in a cost-effective manner for customer delight. Technology leaders play an important role in leading the evolution of business systems, understanding the challenges, and building capabilities to continuously innovate. They should also look outside their organization and find opportunities for collaboration with teams to ideate and innovate. Furthermore, they ensure service quality by implementing process improvement models with the help of widely adopted frameworks to help deliver high-quality services, addressing a vast majority of customers. CMMI, GEIT, and ITIL are a few of the frameworks that can help in adopting practices around service delivery, development, governance, and operation. The adoption of these frameworks and their practices varies widely in the industry. Smaller organizations can start small and focus on a few of the relevant practices, while the enterprises that may want to achieve a certain level of maturity can explore the entire model.

What architects must know

The role of an architect is to understand the need for changing business drivers and technology and their impact on software architecture. They work closely with the business stakeholders to identify new requirements and formulate plans to accommodate those changes to drive business value. As change is the only constant, an architect should embrace the architectural modularity of software systems to be evolved in isolation. They need to align architecture decisions (trade-offs) with the business priorities and understand their impact on business systems. The architect should demonstrate leadership, mentorship, and coaching across teams to facilitate change. They have the responsibility of enabling software teams by providing the necessary learning opportunities to acquire the necessary skills to deliver effectively. These learning opportunities are not only limited to technology training and certifications, but special emphasis should be placed on understanding business systems, business domains, and change management. When it comes to agile teams, the role of an architect is usually played by the senior engineers or the development managers.

In practice, it's essential to understand both the business domain and data. These architects should collaborate with business users to understand the business outcomes and emphasize domain-driven design. They should also analyze how different parts of the application are consuming different datasets, to remodel them as microservices. Understanding the usage patterns of the data allows us to either build or choose the right services that support business needs.

The role of the product manager, product owner, and scrum master

Product managers are responsible for crafting the vision of a product by understanding customer needs, business objectives, and the market. They sit at the intersection of business, technology, and user experience, while the product owners work closely with the stakeholders to curate detailed business requirements for the product or feature. They are responsible for translating the product manager's vision into actionable items to help cross-functional teams build the right product. The scrum master is the coach that acts as a mediator between the development team and product owner to ensure that the development team is working in alignment with the product backlog. They foster an environment of high performance and continuous improvement across the team.

So far, we have learned about the role of leadership in shaping the culture to adopt change. Next, we will discuss the importance of setting priorities before starting the microservices journey. Leadership should keep revisiting these priorities to ensure their alignment with business objectives.

Defining core priorities for a business

Since the microservice architecture is gaining momentum, more and more organizations have started thinking about adopting a new way of building applications. These organizations are setting themselves up for a journey. Startups are challenged with limited resources and deciding what to build in-house and what to outsource. Large organizations are looking to adopt new ways to deliver value faster to their end customers. They are building new applications and also looking to transform their legacy applications into microservices to gain the benefits of being more nimble and agile. As they progress through their journey, they will need to make decisions and deal with lots of trade-offs.

To start, it's important to outline the charter based on your business priorities. If the focus is on building new features and increasing team velocity, then innovation should be prioritized over reliability and efficiency. The following diagram depicts different dimensions that need to be prioritized based on business priorities:

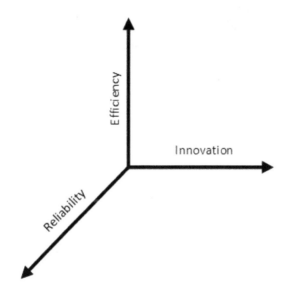

Figure 1.2 – Different dimensions of business priorities

If providing a reliable service to your customer is your primary goal, then reliability should be prioritized over innovation and efficiency. If efficiency is more important, then you must focus on making the right changes to become an efficient organization. By doing so, innovation and reliability may need to be addressed later in the value chain. For example, being more efficient enables experimentation at a low cost, which allows teams to try out new things more often.

Technology leaders need to make sure that their team is ready, and they must be aware of cultural change, new ways of communication, learning new tools and technologies, autonomy in making decisions about their service domains, and be open for collaboration to ensure governance. Technology leaders should bring clarity by defining and communicating the right architecture to the teams.

As an architect, you need to make sure that you are capturing all the decisions that are mutually agreed upon, their benefits, and their trade-offs. You are also responsible for documenting and communicating the implications of these decisions. Documenting these decisions will help you look back and find reasons for making those decisions. This will allow you to make more informed decisions in the future. Identifying good candidates for microservices and modeling them around business capabilities is crucial. You should start with services that are not critical to the business and evolve your architecture incrementally.

Building a new application or transforming a legacy application into microservices is a crucial decision for an organization. It's highly unlikely that you should lay out a detailed plan for such initiatives, but the new services should be designed in a way that they should cater to the current and emerging business requirements. However, understanding the current system can help in identifying areas that can deliver the most impact. Another important aspect is observing the usage patterns of different parts of the application to understand how clients and different parts of the application are interacting with each other. For many organizations, modernizing legacy applications and adding new features to address customer needs are competing priorities. There are mainly three approaches that are adopted as teams start converting those applications into microservices:

- Replacing existing functionality with new microservices that are being built using new tools and frameworks to avoid any technical debt. This requires a huge investment upfront to fully understand the legacy system. It's usually recommended for small legacy systems or a module in a large legacy application that offers limited functionality.

- Extracting an existing functionality that exists in different modules as microservices. This approach is recommended for legacy applications that are well maintained and structured throughout their life cycle.

- As an application continues to grow, it's highly unlikely that the initial design principles are followed during its lifespan. In such cases, a balanced approach should be adopted to allow teams to refactor old applications. This helps them clearly define module boundaries, which can later serve as a means of extracting code from microservices.

The microservice architecture is an implementation of distributed computing, where different microservices make use of parallel compute units to achieve resiliency and scale. Over the last decade, cloud providers have been enabling organizations to reap the benefits of distributed computing by eliminating the need for heavy investment upfront. High-performing organizations are focusing on embracing cloud-native development to accelerate their microservices journey.

Scalability and availability are no longer an afterthought for organizations for building enterprise-grade applications. Cloud-native technologies are playing an instrumental role in enabling organizations to incorporate these characteristics while spanning different clouds environments (public, private, or hybrid). Some of the technologies that are contributing to this development are DevOps, microservices, service meshes, and **Infrastructure as Code (IaC)**.

Cloud-native applications are built to make use of cloud services to provide a consistent experience for developing and operationalizing different environments. This approach has allowed organizations to consume **Platform-as-a-Service (PaaS)** compute infrastructure as a utility and outsource the infrastructure management to cloud vendors. These services are responsible for provisioning the underlying infrastructure, its availability, scalability, and other value-added services. One huge drawback of directly using cloud services without any intermediate abstraction (in the form of a library or framework) is that it makes it harder for organizations to move to different solutions or different cloud providers. **Dapr**, which stands for **Distributed Application Runtime**, uses a sidecar pattern to provide such abstraction between microservices and cloud services. Dapr will be covered in detail in *Chapter 3, Microservices Architecture Pitfalls*.

The cloud-native approach focuses on adopting four major practices:

- **DevOps** is the union of people, processes, and products to enable continuous delivery of value to our end users (as defined by Donovan Brown).

- **Continuous delivery** is a software development discipline where you build software in such a way that the software can be released to production at any time (as defined by Martin Fowler).

- **Microservices** is an architecture style that allows applications to be built as a set of loosely coupled, fine-grained services. These services have logical and physical separation, and they communicate with each other over a network to deliver a business outcome.

- **Containers** or **containerization** is a new way of packaging and deploying applications. The container technology enables packaging applications and their dependencies together to minimize discrepancies across environments.

Software methodology plays an important role in building an application that follows the best practices to deliver software. Twelve-factor app is a widely adopted methodology for building microservices. Let's look closely at the various factors that are essential to understand before starting this journey.

Using the twelve-factor app methodology

The twelve-factor app methodology provides the guidelines for building scalable, maintainable, and portable applications by adopting key characteristics such as immutability, ephemerality, declarative configuration, and automation. Incorporating these characteristics and avoiding common anti-patterns will help us build loosely coupled and self-contained microservices. Implementing these guidelines will help us build cloud-native applications that are independently deployable and scalable. In most cases, failed attempts at creating microservices are not due to complex design or code flaws – they have set the fundamentals wrong from the start by ignoring the widely accepted methodologies. The rest of this section will focus on the 12 factors in light of microservices to help you learn and adopt the principles so that you can implement them successfully.

Code base

The twelve-factor app methodology emphasizes every application having a single code base that is tracked in version control. The application code base may have multiple branches, but you should avoid forking these branches as different repositories. In the context of microservices, each microservice represents an application. These microservices may have different versions deployed across different environments (development, staging, or production) from the same code base. A violation of this principle is having multiple applications in a single repository to either facilitate code sharing or commits that apply to multiple applications, which reduces our ability to decouple them in the long run. A better approach is to refactor and isolate the shared piece of code as a separate library or a microservice. The following diagram shows the collaboration of different developers on a single code base for delivering microservices:

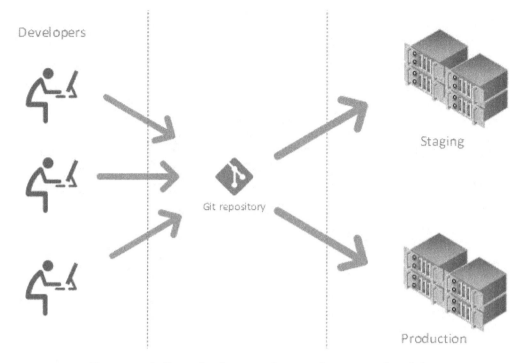

Figure 1.3 – Different developers working together on a single code base

In the previous diagram, a microservice team is working on a single code base to make changes. Once these changes are merged, they are built and released for deployment to different environments (**staging** and **production**). In practice, staging may be running a different version than production, although both releases can be tracked from the same version control.

Dependencies

In the Twelve-Factor App, the dependencies should be isolated and explicitly declared via a dependency declaration manifest. The application should not depend on the host to have any of its dependencies. The application and all its dependencies are carried together for every deployment. For example, a common way of declaring Node.js application dependencies is `package.json`, for Java applications, it's `pom.xml`, and for .NET Core, it's `.csproj`. For containerized environments, you can bundle the application with its dependencies using Docker for deployment across environments. The following diagram depicts how application dependencies are managed outside the application's repository and included later as part of container packaging:

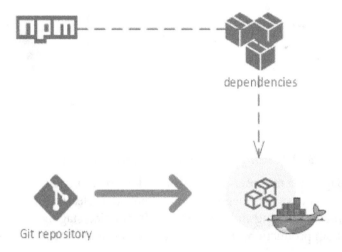

Figure 1.4 – Application dependency isolation and its inclusion via a Docker file

Containers help in deploying applications to different environments without the need to worry about installing application dependencies. **npm** is a package manager that's responsible for managing packages that are local dependencies for a project. These packages are then included as part of container packaging to build container images.

Config

A single code base allows you to deploy your application consistently across different environments. Any application configuration that changes between environments and is saved as constants in the code base should be managed with environment variables. This approach provides the flexibility to scale an application with different configurations as they grow. Connection strings, API keys, tokens, external service URLs, hostnames, IP addresses, and ports are good candidates for configs. Defining config files or managing configs using grouping (development, staging, or production) is error-prone and should be avoided. You should also avoid hard-coding configuration values as constants in your code base. For example, if you are deploying your application in Azure App Service, you can specify configuration values in the Azure App Service instance rather than inside the application's configuration file. The following diagram demonstrates deploying a container image to different environments with different configs:

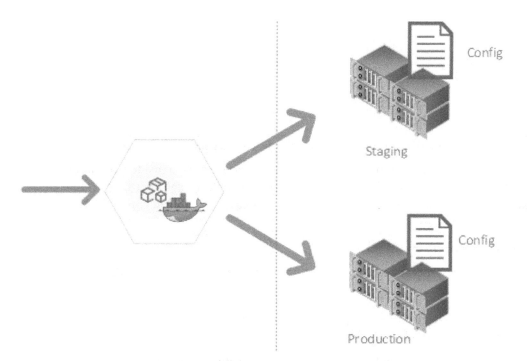

Figure 1.5 – Application deployed to different environments with different configs

A container image represents the template of a microservice and is compromised of application code, binaries, and dependencies. The container image is then deployed to the production and staging environments, along with their configurations, to make sure that the same application is running across all environments.

Backing service

A backing service is a service dependency that an application consumes over a network for its normal operation. These services should be treated as attachable resources. Local services and remote third-party services should be treated the same. Examples include datastores (such as Azure SQL or Cosmos DB), messaging (Azure Service Bus or Kafka), and caching systems (Azure Redis Cache). These services can be swapped by changing the URL or locator/credential in the config, without the need to make any changes to the application's code base. The following diagram illustrates how different backing services can be associated with a microservice:

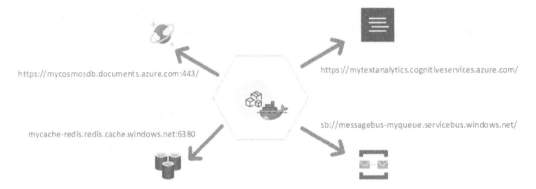

Figure 1.6 – How an application can access and replace backing services by changing access URLs

This diagram depicts a running instance of a microservice hosted inside a container. The microservice is consuming different services that are externalized using parameters. Externalizing backing services helps microservices easily replace backing services.

Build, release, and run

The deployment process for each application should be executed in three discrete stages (build, release, and run), as follows:

- The **build** stage compiles a particular version of the code base and its assets, and then fetches vendor dependencies to produce build artifacts.

- The **release** stage combines the build artifacts with the config to produce a release for an environment.

- The **run** stage runs the release in an executing environment by provisioning processes, containers, or services.

As a general practice, it's recommended to build once and deploy the same build artifact to multiple environments. The following diagram demonstrates the process of building and releasing microservices that are ready to be run in different environments:

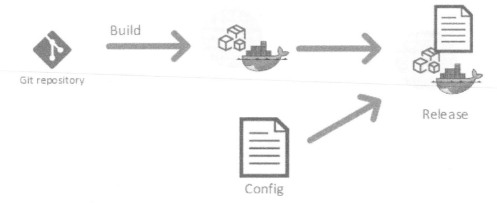

Figure 1.7 – Creating a release for deployment on an environment

Processes

The process is an execution environment that hosts applications. These processes are stateless in nature and share nothing with other processes or hosts. Twelve-factor app uses stateful backing services such as datastores to persist data to make it available across processes or in the case of process failure. Sticky sessions, or storing data in memory or disk that can be reused for subsequent requests, is an anti-pattern. Session states should be externalized either to a database, cache, or any other service. Externalizing the state to a host would be an exception and considered an anti-pattern as it significantly impacts the ability of these processes to run on individual hosts, and it also violates the scalability guidelines of the twelve-factor app methodology.

An example of a stateless process and its stateful backing services, such as Azure Cosmos DB and Azure Redis Cache, is shown in *Figure 1.6.*

Port binding

Traditionally, web applications are deployed inside web servers such as IIS, Tomcat, and httpd to expose functionality to end users. A twelve-factor web app is a self-contained application that doesn't rely on running instances of a web server. The app should include web server libraries as dependencies that are packaged as build artifacts for deployment. The twelve-factor web app exposes HTTP as a service binding on a port for consumption. The port should be provided by the environment to the application. This is true for any server that hosts applications and any protocol that supports port binding. Port binding enables different apps to collaborate by referring to each other with URLs.

Concurrency

The loosely coupled, self-contained, and share-nothing characteristics of the twelve-factor app methodology allow it to scale horizontally. Horizontal scaling enables adding more servers and spawning more processes with less effort to achieve scalability at ease. The microservice architecture promotes building small single-purpose and diversified services, which provides the flexibility required to scale these services to achieve better density and scalability. There are various managed services available on Azure that provide you with a better capability for configuring auto-scaling based on various metrics. The following diagram demonstrates the scalability patterns of different microservices:

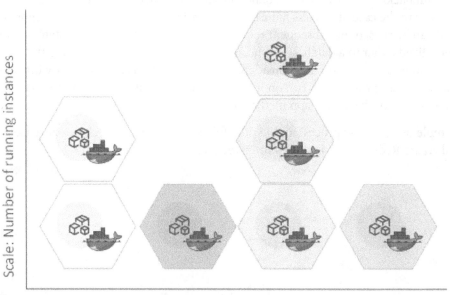

Figure 1.8 – Scalability of different microservices

At a given instance, a different number of microservice instances are running to meet the user's need. The orchestration platform is responsible for scaling these microservices to support different consumption patterns, which helps with running microservices in an optimized environment.

Disposability

The twelve-factor app methodology should have a quick startup, a graceful shutdown, and be resilient to failure. Having a quick startup time helps with spawning new processes quickly to address spikes in demand. It also helps with moving processes across hosts and replacing failed processes with new ones. A graceful shutdown should be implemented with SIGTERM to allow a process to stop listening to new requests and release any resources before exit. In the case of long-running processes, it's important to make sure that operations are idempotent and can be placed back on the queue for processing. Failures are bound to happen; what's important is how you react to those failures. Both graceful shutdowns and quick startups are important aspects of achieving resilience.

Dev/prod parity

"But it works on my machine."

– Every developer

The twelve-factor app methodology advocates having parity between development, production, and any environment in-between. Dev/prod parity aligns with the shift-left strategy, where issues are identified at an earlier stage to allow more time to resolve them with the least impact. It also helps with debugging and reproducing issues in the development and staging environments. A common practice is to use different database versions in development and production environments; this is a violation. In a containerized environment, make sure that you are using the same container image across environments to maintain parity. It's important to have the capability to recreate the production environment with the least amount of effort to achieve parity, reproducibility, and disposability. DevOps practices are crucial for achieving dev/prod parity, where the dev and ops teams work together to release features with continuous deployment. They use the same set of tools in all environments to observe their behavior to find issues.

Logs

The twelve-factor app methodology treats logs as event streams. It never concerns itself with how these events are routed, processed, or stored. In a traditional monolithic application, logging libraries are used to allow applications to produce log files. Cloud-native development differs from traditional development, where the application is built as a microservice. A single request may be processed by multiple microservices before it can produce a meaningful result. Monitoring and tracking the collaboration is important to provide observability of the overall system. The distributed nature of the system makes logging a daunting task that should be handled in a unified manner to support consistency and scalability. The twelve-factor app (microservice) doesn't contain a logging library as its dependency; instead, it uses a logging agent as a separate process to stream logs. For example, these events can be sent to Azure Event Hub for long-term archival and automation, or Application Insights for monitoring or raising alerts.

Admin processes

Design admin/management processes with the same rigor that you use to develop and run any other application. It should run in the same environment alongside other processes while following the same practices discussed earlier as part of the twelve-factor app methodology. Admin processes should be maintained inside the code base, with the config released to environments using the build, release, and run practice. Database migration, running ad hoc scripts, and purging data are all examples of admin processes.

The twelve-factor app is one of the most widely adopted methodologies for building cloud-native applications, and it aims to help architects and developers follow the procedures that are aligned with best practices and standards, especially while building microservices. In the next section, we will explore a few additional factors that are essential for building cloud-native applications.

Additional factors for modern cloud-native apps

In *Beyond the Twelve-Factor App*, Kevin Hoffman discussed three additional factors that are important for building cloud-native applications in the modern era. Let's take a look.

API first

In the API first approach, applications are designed with the intent to be consistent and reusable. The API description language helps in providing a contract that dictates the behavior of an API. For example, Swagger provides the tooling for developing APIs with the OpenAPI specification.

Telemetry

In a cloud-native environment, teams have less control over the execution environment. They are dependent on the capabilities provided by the cloud vendor. Installing debuggers and inspectors is not an option, which makes real-time application monitoring and telemetry a challenge. To address telemetry, you need to make sure that you are collecting system/health data, application performance data, and domain-specific data so that you have complete visibility of the system in near real time.

Security

Security is an important pillar of a well-architected cloud-native application. The cloud-native application is built to run in multiple data centers to serve customers across the globe. They need to ensure that the clients accessing these applications are authenticated and authorized. These applications should identify every requestor and grant access based on their roles. At times, these requests are encrypted using **Transport Layer Security (TLS)** to ensure data privacy while data is in transit. Teams also need to think about securing their data at rest with the capabilities available in different storage platforms. As an example, Azure Cosmos DB and an Azure SQL server allow us to encrypt data at rest.

The preceding 12 factors and the three additional factors we've looked at here have given us a thorough understanding of building cloud-native microservices with an API first mindset. In the cloud-native world, security and telemetry are no longer an afterthought. Approaching security with a zero-trust mindset is essential, while continuously collecting data to monitor system health gives you the ability to make the necessary adjustments to achieve higher availability for the systems.

Summary

In this chapter, we learned about various aspects of setting the mindset for starting your microservices journey. We dissected the microservice architecture into its components and understood the design principles for building microservices. Adopting microservices has several benefits, which essentially help organizations bring business value faster to their end users. We also discussed the challenges of adopting microservices and how these can be addressed.

Later, we discussed the role of leadership, where technology leaders have the responsibility of bringing clarity to drive the overall initiative. Technology leaders bring the right level of focus to empower teams by enabling autonomy and invest in their growth. Architects work with stakeholders to address business requirements to shape the architecture. Finally, we discussed how defining the core priorities for businesses can affect different architecture decisions as you move through your microservices journey. We also discussed the importance of going cloud-native and adopting the twelve-factor app methodology to build microservices. With this knowledge, you can start planning for the adoption of microservices by investing in the right areas to build organizational capabilities.

In the next chapter, we will discuss the role of understanding domain-driven design, anti-patterns, and how to address those anti-patterns. Moreover, we will learn about the importance of domain-driven design, common pitfalls, and some best practices that are essential when building a microservice architecture.

Questions

1. What are the design principles of microservices?

2. What are the architecture components of microservices?

3. How can you build highly scalable, maintainable, and portable microservices so that they're cloud native?

Further reading

For more information regarding what was covered in this chapter, take a look at the following resources:

- *Team Topologies*, by Matthew Skelton and Manuel Pais

- *Microservice Patterns and Best Practices*, by Vinicius Feitosa Pacheco, Packt Publishing

- *Building Microservices with .NET Core*, by Gaurav Kumar Aroraa, Lalit Kale, and Kanwar Manish, Packt Publishing

2
Failing to Understand the Role of DDD

It could be argued that an anti-pattern when building microservices is not having a deep understanding of the business domain or the problem space and then trying to implement microservice tactical patterns. This can lead to chaos or a large, distributed monolith that is tightly coupled and overly complex.

Microservices are collections of services that are designed to meet a common goal or serve a particular business feature. Microservices must work together to meet this common goal. In some cases, a service may be completely unaware of its role when it comes to the bigger picture or may not have knowledge of any other microservices. On the other hand, the teams that build these services should understand the bigger picture and how each service helps achieve the greater goal of the system as a whole or where their service fits in the composition of a larger application.

In this chapter, we will re-examine the characteristics and properties of microservices and then see what role **domain-driven design (DDD)** has in relation to building microservices. We will examine the following topics on our journey:

- What are microservices and how do they align with DDD?
- Benefits and challenges of microservices as compared to DDD
- Lack of team alignment
- Lack of governance
- The whole is greater than the sum of its parts
- Microservices are based on science
- Complex or not complex—that is the question
- DDD
- Lack of awareness of stakeholder needs
- Bounded contexts

By the end of this chapter, you will have learned about the importance of understanding DDD to build cohesive services with bounded contexts. You will also learn about the importance of team building, governance, and awareness of stakeholders.

What are microservices and how do they align with DDD?

The more we describe microservices, the more it sounds as though we are talking about DDD. So, let's take another look at the definition of microservices and see the relationship between what is called microservices and DDD. We will explore the guidelines around DDD and how these guidelines help us in building a microservices architecture that meets the needs of a business.

If you are building microservices, you need to understand what they are, and just to make sure that we are all on the same sheet of music, let's examine the characteristics of microservices. Let's explore what they are, and we will then see how DDD will guide us on our journey in building effective microservices that meet the needs of a business. We will examine the pros and cons of microservices and see how DDD helps us achieve some of these pros and address some of the cons as well.

Something I have noticed in my studies of microservices that has always puzzled me when some folks describe microservices in articles and blogs, as well as in some books about microservices, is that you see very little mention of how much DDD plays a role or influences the creation of microservices and how it really serves, in my humble opinion, as a blueprint or the basis of microservices; it was around before microservices and has kind of gotten the cold shoulder as far as I am concerned.

The DDD and microservice concepts were introduced around the same time and share many of the same characteristics, such as single responsibility, a single code base, separated databases for each service, team alignment, and boundaries.

Eric Evans' book *Domain-Driven Design* came out in the 2003-2004 time frame, and soon after microservices in 2005, Peter Rodgers introduced the term *Micro-Web-Services* during a presentation at the *Web Services Edge* conference. As you study these concepts, you'll see many similarities that these two concepts share.

Let's now explore the similarities between microservices' characteristics and DDD and how they align. First up, we will look at autonomous microservices and DDD's bounded contexts. We will also look at other microservices and DDD characteristics and compare them.

Autonomous

Microservices are small single-responsibility applications that are loosely coupled with one another. A single team of developers can build, test, deploy, and maintain a single small microservice. Each microservice owns a separate code base that a small team of developers can manage. This concept is described in *Domain-Driven Design* when the book addresses characteristics of a bounded context, which has its own code base and is aligned with a single team of developers.

Independently deployable

Microservices can be deployed independently. This frees up a team to update a microservice without rebuilding and redeploying the entire application. DDD supports this concept as well, as each domain and supporting domains generally all have their own code base and adhere to the SOLID design's **single-responsibility principle** (**SRP**). So, when something in one service changes, we should not have to deploy other services at the same time due to that change in another microservice.

Data ownership

Microservices are responsible for persisting the state of their data, and they are also responsible for communicating with other services so that the other service can update the state of its data. This is a change from the traditional single database model, where a separate data layer handles the saving of data. DDD uses aggregates as a concept of transactional boundaries to help use design or data concurrency and consistency. Aggregates should enforce business invariants.

Communication

Microservices communicate with each other by using an interface over **HyperText Transfer Protocol (HTTP)**, using things such as **Google remote procedure call (gRPC)** or **REpresentational State Transfer application programming interfaces (RESTful APIs)**, or via a messaging service. The internal implementation details of each service are encapsulated or hidden from other services. DDD not only addresses service-to-service communication but also communication between teams. It gives clear guidance on how teams interact and integrate with each other. DDD provides strategies for successful communication patterns to assist in building effective microservices that serve the greater good or achieve a common goal of the system. As for service-to-service communication and translation between services, we use a concept from DDD called an open host, which is responsible for the translation of one model to another, like an adapter pattern. In reality, an open host is nothing more than a RESTful web API. As opposed to a RESTful API and making direct services to services calls, I would highly recommend avoiding direct synchronous calls when possible and implement asynchronous messaging services or queues to decouple services from each other. There are use cases where you may need to make direct calls from one microservice to another microservice, but this should be avoided if possible. These direct calls can become an anti-pattern if you rely too heavily on direct service-to-service synchronous calls. This can lead to tightly coupled microservices with direct service-to-service interdependencies, sometimes referred to as distributed monoliths.

Technology-agnostic

Microservices don't need to be built using the same technology stack, libraries, or frameworks. You are free to write your services in any language that you like. DDD isn't concerned with what technology or language you use. In Eric Evans' book, he used **object-oriented programing (OOP)**, and since then, we have witnessed successful implementation using different program paradigms (such as functional programming) that have delivered valuable solutions using the DDD concept.

Gateway pattern service discovery

An API gateway is the entry point for clients. Instead of calling services directly, clients make calls to the gateway, which intelligently routes the call to the appropriate backing services.

Advantages of using an API gateway include the following:

- It decouples the client from backing services. This allows services to be versioned and refactored without needing to update other services or clients.

- Services can use messaging protocols such as the **Advanced Messaging Queuing Protocol (AMQP)** for queues or **publish-subscribe (pub-sub)** scenarios.

- An API gateway such as NGINX that operates on **Layer 7 (L7)** can perform other cross-cutting concerns such as authentication caching, logging, **Secure Sockets Layer (SSL)** termination, and load balancing.

Many of the benefits of microservices are the same in DDD. Let's examine a few of them to compare the similarities.

Benefits and challenges of microservices as compared to DDD

There are various benefits offered by microservices in comparison to DDD. Let's have a look at them, as follows:

- **Highly agile**: Due to the fact that microservices are deployed independently, this makes managing fixing bugs and adding features much easier. You make changes to a microservice without redeploying the entire application, or if something goes wrong, we can roll back an update to a previous version during a deployment. In monolithic applications' deployments, if a bug is discovered during deployment, it can block the entire release, requiring the rolling back of the release. This can cause a delay in the deployment of new features. This trait of microservices being independently deployable is shared with DDD rules around how a bounded context is deployed, and they have their own code base. So, you can clearly see that by understanding DDD, you understand how microservices should be built and deployed.

- **Small teams**: Small teams that are focused on one or multiple microservices. A microservice should be aligned with a small team that can build, test, and deploy the microservice. A smaller team size promotes greater agility. Large teams tend to be less productive and harder to manage because communication is slowed down, management overhead goes up, and agility diminishes. These same principles apply to a domain-driven bounded context as small teams are aligned to a bounded context. As we examine more about microservices, we start to see the one-to-one mapping of the same principles between microservices and DDD. In most cases, teams don't just manage and control one microservice or bounded context but are responsible for and have ownership of multiple microservices, which share a common agenda or business feature.

- **Small code base**: In a monolithic application, it is very common over time for code-base dependencies to become a tangled web or a big ball of mud. Adding a new feature or making changes becomes risky and requires changing code in a lot of different places throughout the code. We should not share a code base or data store; a microservices architecture minimizes dependencies, and that makes it easier to add new features. DDD calls for each bounded context to have its own code base and be aligned to a single team.

- **Mix of technologies**: Development teams can pick whichever technology or programing language best fits their service or their skill set, using a mix of technology stacks as appropriate. This concept is shared with DDD, as each bounded context can use whichever technology or programming language you want.

- **Data isolation**: It is much easier to manage schema updates because only a single microservice is affected by any changes. In a monolithic application, updates or changes to the scheme can become very risky and challenging because different parts of the application may all use the same data or scheme, making any alterations to the schema risky. DDD shares this trait, as well as each bounded context, has its own data, and it is responsible for it.

- **Complexity**: A microservices application is complex when compared with an equivalent monolithic application. Each service has a single responsibility, but the entire system as a whole is much more complex when implementing a solution with microservices. We need to use DDD to help gain deep domain knowledge to distill this complexity down into a model that we can use to build our microservices.

- **Skill set**: DDD and microservices require a higher level of skills, so you should carefully evaluate whether the team has the skill, experience, and motivation to be successful.

We have considered some of the benefits and challenges we face with microservices, and even discussed how some of those challenges and benefits revolve around teams (things such as the code base and skill sets), but let's really hammer home the importance of team alignment, or a lack thereof.

Lack of team alignment

A general rule of microservices is that one team must be aligned or made responsible for a microservice. Ideally, this will be one team per service, but this is not always possible due to limited resources. However, a team can have responsibilities for multiple services, which is common. One team may be responsible for multiple services, but a service should never have multiple teams working on it. One team must have responsibility for the analysis, development, and deployment of a service.

As with most rules, there are exceptions, and there are some rare cases where this may not be possible and where we may need two teams to share responsibility for a service, and teams will have to work together and share responsibility for a service due to organizational or other constraints. DDD calls this a shared kernel, and this requires a high level of communication between the teams sharing the service. We will explore shared kernels in depth later on in the *DDD* section of this chapter.

However, having one team per service keeps teams less dependent on each other and allows them to develop and deploy services at their own pace without relying on other teams. We also avoid other teams having merge conflicts and integration issues. Aligning a single team to a code base relieves many of the issues around testing, development, and deployment.

Now we have examined the benefits and challenges and how teams are aligned to autonomous services, we also need to look at how microservices must work together to form the system as a whole.

Lack of governance

Governance of standards is an important part of any organization. DDD provides a framework for governance—for example, prescribing OOP and clean architecture drawing on good SOLID design principles such as single responsibility to dependency **inversion of control** (**IoC**) are just a few we could name. Governance is all about standards, rules, and guidelines around code quality and architecture. Having solid standards is important but they should not be so constraining that they inhibit innovation, instead having clear standards that will help avoid issues that occur when there are no clear standards.

Governance can be something simple, such as naming conventions, to something complex, such as architecture patterns or limits on resources. So, while we implement coding standards, they should be aligned across the organization and on individual teams to allow flexibility but they should also provide clear communication on what is acceptable and what is not. This can be simple, such as how you name things, to which architecture should be used for a particular context, and context is important when it comes to governance. A rule for one team will not work for another. So, rules and guidelines should be aligned with the context and scope; for example, one microservice may have different governance than another service. That being said, we also need rules that cover the entire scope of a project or an application. Providing clear rules and standards at each level allows for teams to develop rules and standards for their teams but also enforce organizational rules and standards. By establishing effective governance, we can ensure everyone is on the same sheet of music and standards are met in accordance with organizational norms.

Now we have established the importance of good standards and governance, we need to explore how to take those standards and build microservices that meet the holistic needs of an application.

The whole is greater than the sum of its parts

To build microservices, we first need to understand a holistic view of our microservices and use a heuristic approach. We need to understand how they are integrated with each other, how they interact with each other, and the relationship between services.

We need to understand what kind of data each microservice requires to perform its responsibility and what the microservice's output needs to be so that it can meet the needs of services or endpoints that depend on the data it provides. It is when we lack an understanding of the bigger picture that things start to fall apart or bugs are introduced. We need to understand what role a microservice plays in the bigger picture.

We need to understand whether a microservice serves complex business needs or if it is providing something more straightforward and less complex, such as meeting some infrastructure needs. For example, microservices can implement some complex business logic or provide simple generic infrastructure concerns such as repositories, messaging, catalog lookup, and cleanup jobs. Understanding the role our service plays as it pertains to the big picture will set us up for success.

When microservices are based on complex business features, they require more careful consideration and analysis. In cases where microservices are responsible for simple tasks such as persisting data or sending and receiving messages, these types of services require less analysis and can be implemented using best practices. This difference in the responsibility needs to be analyzed and understood so that all the services can work together to meet the needs of the business feature or serve a business domain.

Microservices combine services together to create a system as a whole but are made up of autonomous independent services. Let's explore their autonomy.

Autonomy

Microservices are autonomous, as a bounded context is in DDD. They are only concerned with their own agenda. Some characteristics to consider are outlined here:

- They may not be aware of other services or even know that other services exist.
- Microservices can be deployed separately without depending on other services.
- Microservices can change the way they want to and evolve according to their needs.
- They are only aware of their responsibilities and behavior and as far as they are concerned, that is all they need to do.
- They know what data they need, and they know what data they need to produce.
- They are not concerned with the responsibilities of other services, at least from their view of the world.

But the truth is, microservices are dependent on the data they receive, and the provided data must meet a service's data needs so that the service can carry out its task and produce an output that it is required to. In some cases, this data can be a complex aggregate, or it may be a simple object with a few properties such as an **identifier** (**ID**) to look up or hold a reference to an external aggregate.

If that incoming data is invalid, the service needs to reject it and not allow that incoming invalid data to corrupt its own internal data. If the data that the services are receiving is invalid, then the microservice will not be able to carry out its responsibilities. Therefore, microservices can have a direct or indirect dependency on another service and the data that dependency produces for them. This is common in the laws of interfaces, where interfaces define and enforce contracts, and contracts must be adhered to and honored in order for things to work.

Understanding these relationships between service and data needs is vital to the success of microservices as a whole. Although microservices are autonomous, they still need to interact with other services. They must present a contract or interface, and to honor contracts with other services, they must adapt and react to other services' output.

What comes first – the data or the events?

It could be argued that the data comes first, as we must understand the data first, and then we can figure out the behavior or operations that need to be executed on that data itself and which event needs to be raised on successful completion. We need to know what data we need so that the next event can occur. Do we need to get the data of an external service, and is the data in the right state? If it isn't in the right state, we need to determine what should happen next or which event will occur next.

In modern application development, we use events to inform us of the successful completion of an operation and what kind of data is needed for an event to occur. Events inform us of completed commands or operations or the failures of those operations so that we can notify another part of the system of an event, and it may or may not act based on previous events or some business policy. So, for an event to be possible, we need to understand the data that is needed for the event to happen and to ensure other events or applications will occur as required. Events are in the past tense; for example, an `Order Created` event is an event that happens after the command to create an order was successful, and for the command to be successful it will need all the required order data to execute a `CreateOrder` command.

We need to understand the state of the data, as in the example with the order, as an order traverses through time and space during the order's life cycle, and the state of the order changes. We need to understand whether the data is coming from different sources and being combined to form an aggregate to meet our service's needs. As in the case of an Order Created event, it is likely that all the data needs to execute a command to create an order that came from different sources, such as a catalog service or a customer service offer such as discounts that could have come from a marketing service or loyalty service. All this data needs to be brought together so that an Order Created event can happen.

We need to determine whether this data compilation is complex, meaning that we do not fully understand how the data should be combined, which operations need to be executed, and what state it needs to be in. We need to determine whether the data is well understood and simple, or whether it is a complicated or complex problem. If we determine that the problem is complicated, we may need an expert to conduct some analysis and to recommend how to handle the data compilation or which operations or behavior need to be executed to get the data into the correct state to meet our needs. If the problem is complex, we need to experiment and probe to figure out what our data needs are and how it is combined into an aggregate. If it is complicated, we can ask an expert to help us. If the data and the problem are simple or obvious and well understood, we can employ best practices to get the data into the right state.

To help understand microservices, I think we need to examine the science behind microservices and complexity theory. So, let's explore complex adaptive systems, as this is how microservices can be viewed. Let's explore the concept and see if you agree.

Microservices are based on science

As we examine the definition of complex adaptive systems, we can see some similarities with many of the concepts of microservices, and even DDD. I challenge you to explore this topic of complex adaptive systems deeper as I believe this knowledge is the foundation of microservices, in my humble opinion.

A complex adaptive system is a dynamic network of system interactions, but the behavior of the whole system may not be predictable according to the behavior of the individual components or nodes. It is adaptive as the individual and collective behavior mutates and self-organizes, depending on the change event or collection of events.

Those systems have relationships with adjacent systems and can interact with each other and may fulfill some need for one another, which may be something that they cannot produce on their own. Understanding those relationships is as important as understanding an individual entity or node itself.

Some systems have no knowledge of other adjacent systems. In other cases, an adjacent system may have knowledge of another and interact directly, while others interact indirectly, but still depending on an adjacent system to meet some need it has.

The following diagram illustrates different autonomous systems; some have relationships that are bi-directional and some have a one-way relationship or dependency:

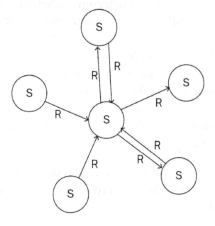

Figure 2.1 – Complex adaptive systems' relationships

In this section, we explored complex adaptive systems. We can think of microservices as being very similar in nature to a complex adaptive system. We now need to determine whether the microservice we are creating is complex or not, as this will help determine our approach for building the microservice in question.

Complex or not complex – that is the question

Microservices can be simple, complicated, or complex, and if we do not understand them, we can fall into chaos. So, we can then use a process—or rather, a framework— to help us to decide whether we are dealing with a simple, complicated, or complex situation in order to determine what type of microservice we are building and what level of effort and resources we will need to build that service, be it people, time, or effort. We can determine whether we can buy something off the shelf for simple situations, or we can implement an existing framework. We can also determine whether we need to involve an expert to tell us which frameworks and tools we need and how to configure such tools. Finally, we can determine whether we are dealing with a complex problem and, if so, we need to assign the best people, resources, and efforts to build our microservice.

The following diagram illustrates that different microservices can address a range of complexity, from simple services such as a catalog lookup to services that are complicated, such as account services or invoicing services, and some services can be complex as they are responsible for implementing business logic or the routing of shipments or goods. We can see that each service can have a dependency on each other, and we now need to determine what level of complexity we are dealing with:

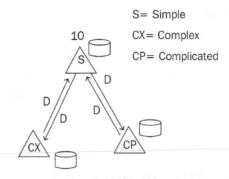

Figure 2.2 – Simple, complex, or complicated problem

I would like to introduce you to a framework that helps in deciding whether we are dealing with a simple, complicated, or complex problem. The framework that will help on this path is the Cynefin framework developed by the scholar David J. Snowden. Let's explore this scientific-based approach.

In the next diagram, we see Snowden's five decision-making domains:

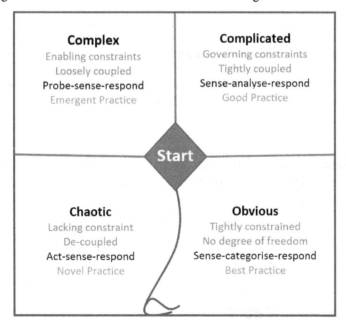

Figure 2.3 – The Cynefin framework helps determine what approach we need to take to solve problems

We will explore each area on the map and see how this may help us in deciding on what type of microservice we need to create and what type of problem the microservice is addressing.

The Cynefin framework contains five decision-making contexts or "domains": simple, complicated, complex, chaotic, and a center of disorder.

By understanding these five domains, we can access how we should approach the creation of our microservices and determine their complexity and which domain they fall into according to the Cynefin framework definition of a problem domain.

Let's examine each of the five domains in greater detail to fully understand how this relates to building microservices.

Obvious contexts – "the domain of best practice"

An obvious context is known as a best-practice domain.

Approach

In this context, the problem is well understood and the solution is clear. We need to Sense – Categorize – Respond, as follows:

- Sense the situation
- Categorize or organize
- Apply an appropriate well-known solution

An obvious context is made up of well-known solutions or processes and does not require an expert to determine the correct course of action.

Risk

The danger with best practices is that teams can try to apply best practices due to previous success when this is not appropriate. Avoid a one-shoe-fits-all mentality when deciding on a solution. Teams can make assumptions that a previous approach will work again. Doing this can lead a team into the chaotic domain, which is illustrated in *Figure 2.3* with the cliff or wave symbol at the bottom between the chaotic and obvious domain borders. To avoid this danger, teams should be willing to try new ideas and listen to suggestions from domain experts and stakeholders.

Snowden's complicated contexts – "the domain of experts"

A complicated context is in the domain of good practice.

Approach

In this context, a problem can have several possible solutions and requires an expert to determine the best course of action. The team understands the questions that need to be answered and who to ask to get answers.

We need to Sense – Analyze – Respond, as follows:

- Sense the situation
- Analyze the situation and ask the right questions
- Apply an appropriate well-known solution based on experts' recommendations

Risk

Teams become too dependent on experts in complicated situations while dismissing creative solutions from the **subject-matter experts** (SMEs) inside an organization. To overcome this, assemble a team of people with experience of a wide variety of levels of seniority, and members of management at all levels in the organization.

Snowden's complex contexts – "the domain of emergence"

A complex context is in the domain of emergent solutions.

Approach

In this context, the problem solution is unknown, and even the questions we need to ask are unknown. A complex domain requires development, analysis, and experimentation. Execute and evaluate to gather knowledge by observing the results of repetitive experimentation and collaboration with domain experts and stakeholders. The goal of a complex domain is to move the problem into a complicated domain.

We need to Probe – Sense – Respond, as follows:

- Probe the problem space
- Sense a solution through experimentation and observation
- Apply an appropriate solution based on knowledge gained through analysis—an experimentation solution is only obvious once it is discovered

Risk

Ensure that the team includes SMEs and business domain experts during the entire process. Accept that failure may be a part of the learning process, and this can be mitigated by using agile development processes with short feedback loops. We start with very little understanding and build our domain knowledge over time until we have obtained rich domain knowledge and can move the solution into a complicated context.

Communication is the key to success. Assemble a wide-ranging group of people, from domain experts to developers, to design creative solutions to solve complex problems. Use brainstorming processes such as event storming to gain a deep understanding of the problem by encouraging business experts and deployment team members to have a conversation, and challenge assumptions to eliminate ambiguity.

A thin line between complicated and complex

Complicated and complex situations look very similar in some ways, and it can be challenging to tell which one of them you're experiencing. One good determining factor can be that if you need to make a decision based on incomplete data, then chances are that you're more likely to be in a complex situation. Complex situations are where DDD shines and where we need to probe the business domain and sense a solution by knowledge crunching and experimenting in an iterated agile way with business domain experts. The following diagram illustrates where DDD fits into the Cynefin framework:

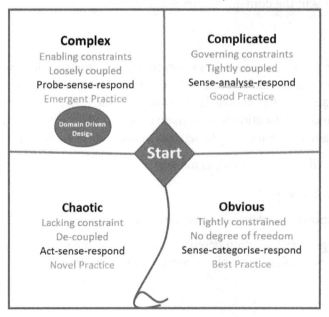

Figure 2.4 – The Cynefin framework where DDD is used to solve a complex problem

Let's move on to the next context of chaos and discuss what should we do in the midst of chaos, which is take full control.

Snowden's chaotic context – "the domain of novel solutions"

A chaotic context is in the domain of novel solutions.

Approach

A domain of chaos is where a problem or situation is out of control and the first order of business is to stabilize the situation or to contain the issue at hand. Many times, this requires a quick fix, which may not be the best solution, but as long as it stabilizes the issue, it is good enough for the time being.

We need to Act – Sense – Respond, as follows:

- Act to implement a quick stabilizing solution
- Sense: assess the situation and determine the next steps
- Respond: take action to move the problem to a complex context

Risk

In order for a chaotic solution to be successful, we must conduct a risk assessment to determine risks and prioritize them. During a crisis, it is important to obtain accurate information by way of effective communication. This information will help you move a problem into a complex domain. Again, as previously mentioned, lack of knowledge can land you in a chaotic domain, so knowledge should be a priority.

Many times, we find ourselves in a disordered space when we start out, as shown in *Figure 2.4*, as disorder is the starting position.

Snowden's disorder contexts – "the space in the middle"

A disorder context is a space in the middle of the starting point.

Approach

A domain of disorder is when you don't know where you are, so you are in disorder.

We need to Assess – Sense – Move, as follows:

- Assess the information you have
- Sense what you know and what you don't know, then gather enough information
- Move to the correct context based on the information you gathered to develop a solution

Risk

In analysis paralysis, you can get caught up in an endless loop of analysis when you need to seek experts to help with the assessment phase.

Our path is clear

Once we have determined whether the problem is simple (obvious), complicated, or complex, we are better informed on which approach we should take to solve the problem associated with that business domain or business feature we are trying to address or solve.

If the problem is simple (obvious), we can use best practices—for example, catalog lookup. If it is a common business problem, we can buy something off the shelf or use a well-defined framework. The point here is that we don't have to write custom code; we can employ some pre-existing tool or process to handle that context and its data and behavior.

If the problem is complicated, we can employ an expert to help implement good practice and suggest which tools or frameworks we need, or a combination of them, to handle the needs of that microservice's responsibilities.

If we determine we are dealing with a complex problem space, we can use DDD "probe sense respond" and discover emergent behavior that evolves from a greater understanding of the problem domain. This can be achieved by using event storming and a DDD model-driven approach, both of which we will explore next. Let's start with event storming.

Event storming

Event storming is a workshop-based methodology that is very effective at helping you to quickly gain an understanding of business workflows and obtain a deeper knowledge of the business domain in order to deploy software that meets the needs of the business. Compared to other methods, it is extremely lightweight and interactive, intentionally requires no computers to be conducted, and focuses on the language of the business. The result is expressed in colored sticky notes: orange for events, blue for commands, yellow for aggregates, light purple for policy, hot pink for issues, green for a read model (view of the data needed by a user to make a decision), and light pink for external dependencies, all of which are placed on a wide wall from left to right along a timeline. The business process is "stormed out" as a series of domain events that are denoted as orange sticky notes with business and development team members present working together, which is referred to as team model building. This requires a facilitator.

Event storming was invented by Alberto Brandolini and is based on DDD. Event storming is a fast way to build a logical model, domain model, workflows, enterprise integration, and enterprise solution requirements. The idea of event storming is to bring together the software development team and domain experts to brainstorm and to learn from each other. To make this learning process more efficient, event storming is meant to be fun. It encourages everyone to participate and there are no bad ideas. Everything posted on the timeline will be challenged and refined as the process moves through its steps. The people invited to the event storm are vital to its success. The team consists of five members from the development team, such as developers, testers, architects, and product owners, and five people from the business from various levels of the organization—from analysts and managers to team members, and even to senior leadership, who have different points of view and deep domain knowledge. The name was chosen to show that the focus should be on domain events and the method works similarly to brainstorming or agile modeling's model storming.

DDD

DDD was introduced by Eric Evans in his blue book *Domain-Driven Design: Tackling Complexity in the Heart of Software*.

Eric Evans describes three key concepts that will help us on our journey of understanding the needs of our business and build microservices that will meet the needs of the business. These three concepts distill our domain down into a model that we can implement and tactical patterns to make the implementation of our microservices successful. Those concepts are listed as follows:

- A lack of awareness of stakeholder needs
- A knowledge-crunching domain model
- **Ubiquitous language** (**UL**)
- Binding the model to the implementation

We will talk about each of these concepts in detail in the next sections.

A lack of awareness of stakeholder needs

The main purpose of developing software is to meet some stakeholders' needs. What if I told you that stakeholders are not just the folks asking for software to be developed? In truth, stakeholders encompass everyone involved in the process, including the development team and the business users. Each stakeholder has different needs—the business stakeholders' needs are different from that of the development team, and all the stakeholders' needs must be met. The business people need an application to meet the business needs or to provide a new feature. The development team needs are around developing the software, and although the needs of each stakeholder are different, they need to be addressed in order to successfully develop microservices. Communication is key to meeting all the stakeholders' needs, and DDD helps us meet all their needs by providing a process or framework for understanding these. The strength of DDD is to understand the problem space that analysis and communication need to occur first. DDD helps tie all these needs together with the business stakeholders' needs, and these are met by developing the domain model. The developers' needs are provided by the code model, which is derived from the domain model. The domain model is developed using UL and knowledge crunching, and then, we bring it all together with an implementation of the code model. We can now examine how DDD addresses stakeholders' needs and provides an awareness that we need to be successful. We will start with knowledge crunching.

A knowledge-crunching domain model

It is vital to heavily involve stakeholders and business domain experts in the creation of a model. In fact, it could be argued that it is impossible to build a model without them. As this may seem a no-brainer, it is not that we don't include the business folks in gathering requirements and documentation. It is the old methods that missed an important concept: that this is never-ending and should not only be done in the beginning. Many times, the level of interaction with business experts and the amount of time we spend with them is inadequate or comes from a single source. I have also experienced where most interactions occur between an analyst or a product owner during the deployment life cycle of an application, and this can introduce a **single point of failure** (**SPOF**).

To illustrate this point, I'd like to use an analogy of a car accident. If you wanted to know what really happened in a car accident, you wouldn't just ask one person if you were responsibly investigating the cause of an accident. You would want to ask all the witnesses, and then you could come to a better conclusion as to what happened or what caused the accident. So, we need deep knowledge of a domain to be successful in developing an application that is made up of microservices, and we need to obtain our knowledge from as many sources as possible from the domain and not just rely on one person. Invite as many domain experts as you can to knowledge-crunching sessions so that you can, as in the car-accident analogy, come to a better conclusion as to what it is you're building and how you should build it.

Or, in upfront design methodologies, interactions with business experts occur in the beginning and may or may not occur again until several months later to demonstrate what the development team has come up with. Many times, these demos result in "this is not what we wanted," or "this is nice, but this is actually what we wanted." Therefore, a team model-building approach should be used when building a domain model. This is vital to your success. The team should communicate often, and this should occur as often as required in a continuous and looping interactive way. When possible, include the business in an agile ceremony such as story refinement communication with the development team, and this should happen often and continue throughout the life of the application. This is especially important in the initial stages and may taper off as the application reaches maturity and enters the maintenance phase. If the application is ever-evolving, as many applications are, then the cycle of communication between the development team must persist and become part of the culture of the organization.

UL

Communication is vital to building effective microservices. Our services should adhere to the language of the business and should reflect this to avoid technical terms and ambiguity.

Domain experts, SMEs, and the people you will need to knowledge crunch with in a business may have a limited understanding of tech jargon and will act as though they do know even if they don't to avoid looking naïve, and nothing could be further from the truth. Believe me—domain experts are very intelligent, and it is their knowledge that you need, not the other way around. In fact, they have a very complex language of their own, and it is we, the developers, who need to quickly learn and understand their language. We must adapt to speak to and understand the business experts, which will become our shared language with the business and will form the UL. Not only are we required to learn the business language, but we also need to implement it in our design by naming our development artifacts such as services, objects, and modules in the language of the business.

Using UL is very important as a project grows and more development teams get involved in building and owning separate domains and developing the microservices for those domains. This growth can create an even bigger need for a shared language, as teams will need to communicate with each other and the business, and we all need to be speaking the same business language. Developers will still have the language they speak to each other, but the names of classes, services, methods, and functions should always reflect business behavior and properties and have a business name. We observe translation being an option, but that leaves room for mistranslations. These translations can happen between business experts and developers, or they can happen between development teams. Translations present a risk as something could always be lost in translation—something we have all heard before.

Domain experts have little knowledge of technical jargon as it pertains to software development; on the other hand, domain experts use the jargon of their field, which can come in various flavors. Developers, as compared to domain experts, may understand and discuss a system in descriptive technical terms, programming jargon, and functional terms, not having any meaning in the business domain experts' language. Developers may also develop abstractions of a model that support their design but these are devoid of any meaning to the domain experts, which creates confusion and hinders communication.

Developers who are working on different parts of the problem work out their own design concepts and ways of describing the domain, thus it is common for communication to break down or ambiguities to arise. Across this linguistic divide, the domain experts vaguely describe what they want in the application. While developers struggle to understand a domain that's new to them, an incomplete understanding shows up as bugs or—worse—completely failing to meet the business need. A few members of the team can manage to become bilingual, but they become bottlenecks of information flow, and their translations are inexact in many cases.

On projects that do not share a common language, developers must translate business domain experts' language to their development language. Domain experts translate between developers' languages and also have to translate from one domain expert working in a different domain to the language of their business domain. Developers even translate between each other. These many translations muddle and dilute the model concepts, which leads to assumptions that lead to the disruptive refactoring of code. This roundabout of communication conceals important formation and details about the model, which leads to team members using terms and definitions differently and being completely unaware of this. This leads to unreliable microservices that don't work together or belong together. The laborious effort of translation prevents the interaction of knowledge and ideas that lead to deep model insights and deep domain knowledge, which is vital to success.

Binding the model to the implementation of our services

Eric Evans tells us DDD uses a model of a domain to solve problems of our business domain, helping us map our logical model of a problem domain to the implementation model of our application. He explains that maps, which are models of the world, are abstractions of reality, and so are models of the domain, which helps us abstract the business reality into a working model that we can implement in software. Through knowledge crunching, a team of developers and domain experts works to distill a torrent of chaotic information into a practical model.

A model-driven design is a connection between a model and the implementation. UL is the mechanism that enables the exchange of information between developers, domain experts, and the software. The result is that the team is able to build software that provides rich functionality based on a piece of deep domain knowledge, and the fundamental understanding of that gives a business a competitive edge, which is the core domain.

Success with a model-driven design is sensitive to detailed design decisions. We need to use the language of the business, speak it out loud, and work together as a team to build the model. It is through the abstractions we dig down into the details, but we must start somewhere. Just like when we plan a route for our family vacation or trip with a map that is a model of reality, we can do the same by creating a map/model of our business domains and workflows to guide us on our journey of building highly effective microservices.

Not a silver bullet

Even though we will not use DDD to solve all our problems, such as in the case of developing in simple and complicated contexts, as we explored in the Cynefin framework, we will discuss simple and complicated contexts during knowledge-crunching brainstorm sessions to identify what is complex and what is not. Then, we can focus most of our analysis efforts and experimentation on complex domains. You will want to put your best people and direct more resources and efforts toward these complex domains, or what is called a core domain in DDD.

Core domains make our business unique, set us apart from our competition, and give us an edge. Core domains are the domains we will focus much of our efforts on, as opposed to complicated and simple contexts, which are supporting domains or subdomains in DDD. Subdomains deal with infrastructure concerns or simple data persistence operations. These subdomains are still important, as we need to understand their role in a holistic view of our problem space or business domain.

It is vital that we have a holistic view of our problem space and understand what core domains and supporting domains are. This will inform us of which services we need to create. Armed with this understanding, we can go about building the type of service we need, whether it is a supporting type of microservice of a subdomain or a core domain microservice with a high level of complexity. The microservices we build need to support the domain model.

DDD is not really one thing, but rather, it is a collection of best practices much like DevOps, which has been described as a collection of people, processes, and tools. DDD was designed by Eric Evans with the agile process and concepts of **continuous integration/continuous delivery** (**CI/CD**) in mind and employs well-known strategic and tactical patterns from the likes of Martin Fowler and Robert Martin. It is based on best practices of **separation of concerns** (**SoC**), clean architecture, and SOLID principles, and includes Martin Fowler's domain module, to name but a few. This collection of lessons learned and best practices that DDD follows will help us to solve complex business problems and distill our business model down into abstractions that employ model-driven design.

Model-driven design is a concept in itself but, as mentioned previously, DDD is made up of best practices, and model-driven design is employed as part of DDD. Model-driven design provides us with a way of translating our business logic model into a software model that we will use to build an application—or in our case, our microservices. By understanding the role that DDD plays, we can build microservices that meet the unique needs of our business domain.

Next, we will look at how we slice up business domains into boundaries. This can be done on a whiteboard and is meant to be informal to facilitate a conversation. Don't worry about perfection at this point—this is more about discovery.

Boundaries are important

Boundaries are not new to OOP, as with the concept of encapsulation, which we use to hide our internal implementation details and only expose them through a well-defined interface. As we move to microservices architectures, we will soon realize the need to have a deeper understanding of those boundaries and how we integrate within the boundaries themselves and with external boundaries. The boundaries represent a different business concept or support a similar concept but operate off a different model with different concerns or responsibilities within the application's own overall boundary, as shown in the following diagram:

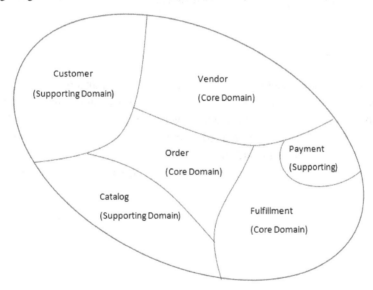

Figure 2.5 – Slicing up the boundaries of an application

As we have discussed before, we build our domain model based on the abstraction of business reality, and from that abstraction, we understand the boundaries of our domains—not just on what's inside those boundaries, but we also need to understand the edges of the boundaries as well as the relationships between those boundaries of our domains. We use the UL we developed by knowledge crunching with the domain experts and the development team to define those contextual boundaries. This leads us to the concept of bounded contexts. This is where we begin the process of aligning our domain logical model to a tactical implementation of the model in code, which will translate into a microservice or a collection of microservices.

A bounded context aligns with the domains we have identified and will form our microservices and establish our boundaries around them.

Bounded contexts

As we build microservices, we align them to a bounded context, and those bounded contexts align with a particular domain model. The bounded context makes up the components of the domain model, from the database scheme to behaviors and domain objects. Each bounded context has its own UL, its own model, and sub-models.

As we can see once again, a bounded context aligns with microservices, as each bounded context can be one microservice or a collection of microservices.

Once we define our bounded context, we need to better understand how they are related, and we can do this with a very loose sketch known as a context map, as demonstrated in *Figure 2.6*.

Context maps

To understand the relationships between our bounded context and how they fit together in the bigger picture, we create a context map. This map is a great communication tool for development teams as it helps facilitate a conversation about technical, tactical, and integration issues. A context map helps us to visualize the lay of the land, understand communication between contexts, and understand how the teams aligned to a different bounded context communicate with each other as well, as shown in the following diagram:

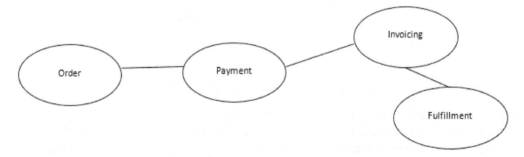

Figure 2.6 – Bounded context in a context map

A context map is not purely a technical point of view but is also how teams integrate and communicate with each other. This communication is vital to successfully develop microservices that meet the needs of the application as a whole. Context maps help us understand which team has a dependency on another team or whether they are upstream or downstream. We will now explore the relationship between teams and their dependencies on other teams and how they interact.

Relationships between teams

Teams working on separate microservices need to be able to develop their microservices in isolation but also need to communicate effectively and integrate with other teams. Next, we will explore those relationships and explain how they work in a context map, as shown in the following diagram:

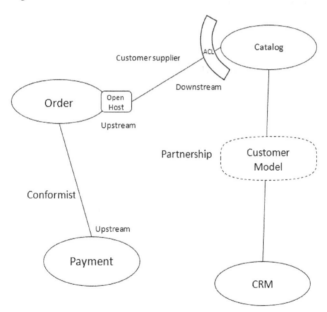

Figure 2.7 – Context map expressing the relationships between a bounded context

Let's discuss the components shown in the preceding diagram, as follows:

- **Customer supplier**: When teams are not working together on a common outcome, to avoid the upstream team from dictating all decisions that can potentially compromise the downstream team.

- **Shared kernel**: When two teams are working on the same application and share it in two separate bounded contexts, they have a lot of interchangeable language and definitions of the main concepts and logic in the bounded context.

- **Collaboration and sharing part of the model to ease integration**: In *Figure 2.7*, for example, Customer Model is a shared kernel and forms a partnership.

- **Conformist**: When an upstream context is not able to collaborate or is out of the control of the organization or is a black box, then the downstream context will have to conform to the upstream context when integrating with those external dependencies.

- **Partnership**: When two teams are working on different contexts but are working toward a common goal, a partnership can be agreed upon to ensure that collaboration on integration between the two contexts can be completed.

- **Separate ways**: When the cost or risk of integration between contexts is too great due to political constraints or complexities, teams need to go their separate ways and not integrate with the system.

Anti-corruption layer

To protect our model, we need to use a translation or adapter layer to map the external service to our internal model to protect our model from getting corrupted by the external service. An anti-corruption layer, as shown in the following diagram, is a one-to-one translation layer that translates one model to another:

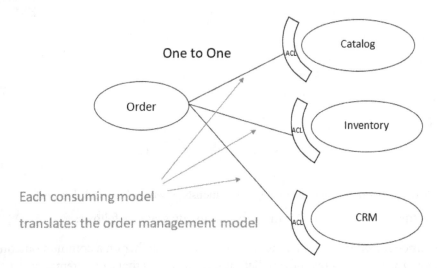

Figure 2.8 – The anti-corruption layer sits on the edge

As endpoints or more adapters are needed per bounded context, we will have increased maintenance and this can become complicated to keep up with, so we can use a single endpoint for all of our adapters, and that is called an open host.

Open host

An open host, as shown in the following diagram, sits in the service layer on the edge and provides translation for multiple endpoints of an external bounding context to protect the internal model from being corrupted. An open host helps when we have many services that need translating and maintenance becomes complex. We can implement an open host with endpoints for each translation:

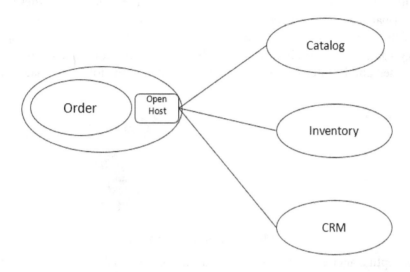

Figure 2.9 – An open host sits on the edge of a service

Microservices expose RESTful APIs to implement an open host, as demonstrated in *Figure 2.9*. These APIs have endpoints that provide access to internals and translations from one model to another.

Aggregates

In DDD, an aggregate can be a confusing concept. One way of looking at it is to think of an aggregate representing a domain concept such as an invoice or an order. Aggregates are a transactional boundary and enforce business invariants by grouping entities related to business rules and invariants together in one transaction.

For example, if an order is shipped before payment is received, then we have broken the invariant or business rule as it makes sense that payment must be received before orders are shipped. In the previous example, we have placed payment in the wrong order and the application is now in a corrupted state as the business invariant is broken.

Transactional boundary

Guaranteeing the consistency of changes to objects across complex business associations can be challenging, to say the least, as we need to enforce business invariants. If our aggregates include common objects such as an address that other aggregates can reference and have different business rules around, these common objects can create contention, leading to locking issues, as aggregates can try to update the same object at the same time. These locking issues negatively impact performance or even lead to crashes of an application. By balanced grouping of our objects based on avoiding high contention and tightening of strict invariants, we can avoid issues with locks and give common objects their own aggregate transaction boundary, which will make a common object responsible for updating itself and its internal objects, as demonstrated in the following diagram:

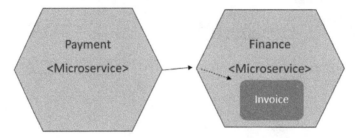

Figure 2.10 – Changes are made through the root

To turn a conceptual aggregate into implementation, we need to follow some important rules for all transactions.

Each aggregate has a root entity that has a global identity, and it also has internal objects that have an internal identity and can only be traversed through the root. No external entity can have a direct reference to an internal member of an aggregate.

All changes to the state of an aggregate and its members are done through the root. If external aggregates need an internal object from another aggregate, they can request a copy from the root of that aggregate. Internal members of an aggregate can have a reference to an external aggregate root using a global identity.

Only the root aggregate can be obtained by directly querying the database. All other objects must be found by the traversal of the association of the root aggregate.

If a delete operation is executed, it must at once remove everything within the aggregate boundary. When the state or change of an object within the aggregate boundary is committed, all invariants of the whole aggregate must be satisfied.

Transactions that change state are hard enough to manage, but when we try to update state and satisfy some request-response design, this can make life much harder than necessary. Let's discuss this by discussing how data flows by direction.

Bidirectional communication between services creates complexity.

Bidirectional communication by using a request-response model can tightly couple services and increase complexity, as demonstrated in the following screenshot:

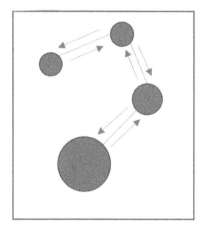

Figure 2.11 – Bidirectional communication

This can often occur when teams align aggregates to **user interface** (**UI**) concerns, which is a common mistake. Aggregates should align with the business rules and domain invariants, as we have discussed previously, so avoid the trap of looking at the world through the eyes of the UI. To simplify this, move to a one-way direction and separate the responsibilities of command and query or request and response. There are situations where this is unavoidable—in the case of a request to return a response, the response may be needed to proceed, as in the case where a UI layer needs the catalog of all available items in order for a customer to select an item to purchase. In the case of placing an order, this will turn into a command and should be a one-way command to create an order with the selected items, persist the order, and raise an event to signify that an order was created so that other services can execute their role in the workflow.

Unidirectional design is a better choice, as we first need to ensure we have enforced our business invariants and that domain objects are in the correct state. This can be achieved by employing the **Command and Query Responsibility Segregation (CQRS)** principle. Commands are a one-way concept, as illustrated in the following diagram:

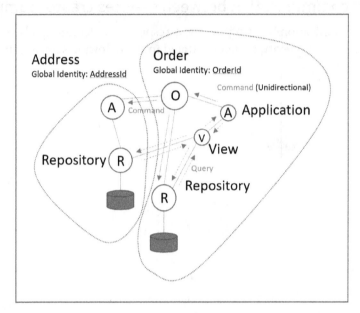

Figure 2.12 – Command query separation

Then, group your aggregates based on the following:

- Business invariants

- Language

- Association and logic grouping supporting the business rules

As shown in the following diagram, we group aggregates and their internal entities based on their transactional boundaries, which enforces the business invariant or business rules:

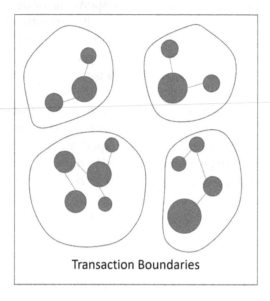

Figure 2.13 – Transaction boundaries grouped by association

Aligned teams

As we have learned some important guidelines around building microservices, one important aspect to consider is the people who make up the team that will build microservices. Some important things to consider are listed here:

- Skill set
- Motivation
- Resources
- Culture

As we have learned, microservices and DDD both state that teams are aligned to a service or a bounded context, which can represent a microservice or a collection of microservices within their respective boundary. Those teams are responsible for developing, testing, integrating, and deploying those microservices in an agile and autonomous way. This means they do not depend on other teams to deploy their service, as they own a separate code base that can be independently pushed and deployed to production. We need highly motivated people to build complex microservices, as the challenges can be considerable and have a steep learning curve. We need to have the time and resources required to build complex services and that also means people, who are the most valuable resource of all. We will want to put our most talented and skilled people on the teams that need to build complex services. We cannot understate the importance of an entire organization embracing a team model-building approach and buying in and supporting this kind of approach. The culture of an organization must evolve to realize the importance of the entire team of business domain experts combined with the deployment team, working closely together during the entire development life cycle.

Iterative process

Short feedback loops will allow for the experimentation and exploration of a domain by deeply involving business domain experts in our agile ceremonies and constantly keeping the domain model in line with the code model. We must leverage UL to remove ambiguity by working closely with business experts. We can evolve our microservices to meet great business goals by gaining deep domain knowledge through our shared language.

Continuous evolution of a model occurs as we knowledge crunch and gain deeper domain knowledge. This allows an application to improve and emerge from the knowledge we have obtained. We do this by using an agile process that employs short feedback loops, as opposed to heavy upfront analysis and design, as in a waterfall approach.

We start with a **minimum viable product** (**MVP**) and iterate on the model. As we gain domain knowledge, the evolution of the model into a well-designed service is realized. Design is still a part of the process—it is just done in smaller chunks and in more effective agile ways.

By understanding DDD's role in building microservices, we are better equipped to build the correct microservices and commit the correct amount of effort and resources in building the services. We now understand that not all microservices are equal and that they have different levels of complexity. We also now understand that microservices and the teams that build them need to integrate and communicate with each other, and DDD gives us the tools to help small teams build, test, and deploy microservices. We have learned the importance of boundaries and how DDD bounded contexts and aggregates help identify those boundaries. In my opinion, an anti-pattern of microservices is not understanding DDD, not only its tactical patterns but also its analytical model-driven strategic patterns and the power of UL.

Summary

We started the chapter with an overview of what microservices are and how they pertain to DDD for you to understand their relationship. We also looked at the benefits and challenges of microservices as compared to DDD.

Later, we learned how DDD relates to the characteristics of microservices and that teams need to be aligned to a service, ideally a one-to-one team alignment to a microservice. Each service should have its own code base managed by one team. We learned that we need to take a holistic view of our microservices architecture. We also discussed the Cynefin framework to determine the type of problem we are trying to solve, in order to understand the amount of effort and resources required to build our microservices.

We talked about how each boundary is based on a logical grouping of aggregates that are transaction boundaries and enforce invariants of the domain. We learned that we may not get this right first time and, as our knowledge of the domain deepens, the boundaries can ebb and flow over time. Lastly, we learned about how microservices communicate and integrate with each other, and also about how teams communicate and integrate.

With this, you are now aware of the importance of understanding DDD to build cohesive services with bounded contexts, as well as the importance of team building, governance, and awareness of stakeholders.

Questions

1. Is DDD important to your success in building a microservice?
2. Which three concepts help in distilling a problem space into a model?
3.   What is the most important thing to understand about aggregates?

Further reading

- *Domain-Driven Design - Tackling Complexity in the Heart of Software.* Eric Evans, Pearson.

- *Event Storming* by Alberto Brandolini (`https://www.eventstorming.com/book/`)

- `https://docs.microsoft.com/en-us/azure/architecture/`

3
Microservices Architecture Pitfalls

These days, many organizations are looking to bring agility as part of their product supply chain and optimize every aspect of their delivery process by incorporating customer feedback. Most of these organizations are already on a journey to adopt the new ways of building applications. DevOps, cloud-native, containers, and microservices are the front-runners in helping these organizations to differentiate themselves from their competitors.

Microservices is an evolutionary architecture where lessons learned from prior architectures are addressed to avoid potential pitfalls. It is important to understand these pitfalls to avoid making the same mistakes while building a microservice architecture and making things even worse. Microservices have many advantages, but careful planning and design is required to reap their long-lasting benefits.

In this chapter, we will evaluate different architecture patterns in the light of microservices design principles to learn how different choices can affect the microservice architecture. To support this explanation, we will use an example of an e-commerce application and apply different architecture patterns to understand their impact.

The following topics will be covered in this chapter:

- Layered architecture and its challenges
- Over-architecting microservices

- Overusing frameworks and technologies
- Not abstracting common microservice tasks
- Lack of knowledge about the microservices platform
- Neutralizing the benefits of microservices by adopting a frontend monolithic architecture

By the end of this chapter, you will be able to understand the motivation behind different architecture styles adopted in microservices, their pitfalls, and their potential solutions. In addition, we will discuss how over-architecting microservices and trying out a variety of tools, technologies, and frameworks can lead to an increase in the complexity and manageability of microservices.

Layered architecture and its challenges

The **layered architecture** is a software architecture approach where applications are organized as a set of components in multiple layers. Each layer contains multiple components addressing a specific area of concern, where each layer can only communicate with its associated underlying layer. This way of communication helps in avoiding cyclic dependencies in a layered architecture. The layered architecture does not specify the number of layers, but it's common to observe four layers (the **presentation layer**, **business logic layer**, **data access layer**, and the **database layer**) in most applications. The layered architecture helps organizations create different teams aligned with different technical capabilities while they build applications.

However, the ability to make changes in a layered architecture is a constraint due to the tight coupling between different layers. As applications grow, making changes becomes more time-consuming due to a lack of understanding of different domains inside the application. Any change needs to be thoroughly analyzed by different teams for its impact before implementation, which increases communication between teams. The deployment frequency of these applications is usually time-consuming and requires building, testing, and releasing the complete application. If different teams are working on different feature branches, then these branches should be merged into the master branch. Deployment in a production environment is usually planned and scheduled in off-hours, which results in application downtime. Compared to DevOps, the deployment cadence of layered applications is usually every quarter.

Microservices give teams the freedom to choose the right architectural style for their implementation, whether it's a layer, tier, or something else. Let's take a closer look at the microservice architecture in terms of layering schemes.

Microservice architecture

The layered architecture is not a bad practice; it allows teams to structure code in different layers to help them understand services. Adopting a layered architecture to implement a microservice could be the right choice as your microservice starts getting heavier. However, in the light of microservice design principles, every team has complete autonomy to choose the right technology for building each microservice. The following diagram is a schematic representation of the microservice architecture for an e-commerce website :

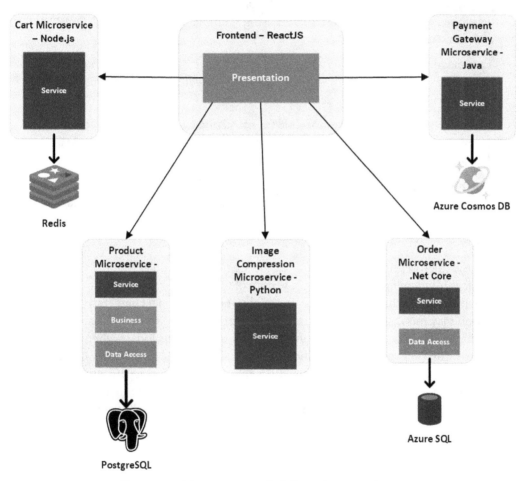

Figure 3.1 – Microservices with different layering schemes

The preceding diagram demonstrates a microservice architecture with different microservices, each with its own layering scheme. Let's discuss each microservice here:

- The **cart microservice** is responsible for managing customer carts while shopping. The service is developed using Node.js and Redis cache using the *service* layer. The service layer contains business logic and data access logic to persist data in the Redis cache.

- The **product microservice** is responsible for managing products. The service is developed using PHP and PostgreSQL using multiple layers (service, business, and data access).

- The **image compression microservice** is responsible for compressing product images for display. It has only one layer that encapsulates all the logic that's necessary for processing images.

- The **order microservice** is responsible for managing customer orders. The service is developed using .NET Core and Azure SQL using multiple layers (service and data access).

- The **payment gateway microservice** is responsible for managing payments. The service is developed using Java and Azure Cosmos DB. The microservice has only one layer that encapsulates all the logic necessary for processing payments and persistence.

In theory, the microservice architecture brings more flexibility and autonomy to microservices by allowing you to choose a variety of technologies. However, it comes with a caveat, where each technology has its own pitfalls and operational characteristics. An organization may miss out on building internal expertise and a knowledge base due to the proliferation of these technologies, and the ramp-up time gets longer when engineers switch teams. Also, it becomes quite difficult to create and have shared library components as various languages are being used in the organization. In practice, organizations prefer to consolidate on fewer technologies. Next, we will explore how designing individual microservices can affect the maintainability, deployability, and scalability of microservices.

Over-architecting microservices

Over architecture or **over-architecting** is a subjective term and has a lot of room for debate but in my opinion, it's the time and effort you put into solving architecture problems you don't have or can foresee, which results in an architecture that's hard to read and follow. As an architect, you should always be striving to make the overall architecture more understandable and maintainable to support future enhancements. Architecting microservices can be overwhelming enough to influence teams to look for similarities across the system and apply generic solutions rather than focusing on use cases. For example, if too much time is spent on over-architecting the application earlier in the project, this may result in precious time being lost, which causes delays when it comes to starting coding activities. A lot of the time, application design evolves as teams become more knowledgeable about the use cases. In the next few sections, we will cover the common pitfalls when it comes to over-architecting microservices.

Entity microservices

Entity microservices is an architectural pattern where microservices are designed to be fine-grained and modeled around a single entity. In most cases, these microservices contain simple **Create, Read, Update, and Delete (CRUD)** methods. An example of such an entity microservice is demonstrated in the form of a cart microservice:

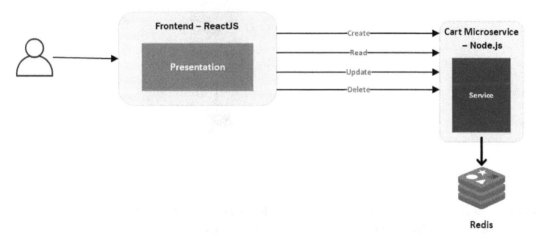

Figure 3.2 – Entity microservice

In the preceding diagram, the cart microservice has been designed as an entity microservice that only exposes CRUD methods. The presentation layer is built using **React.js**, which can invoke methods that are available in the cart microservice. The drawback of using an entity microservice is its limited functionality to address a variety of business use cases. In this scenario, the consumers of the entity microservice need to implement additional capabilities to address business use cases related to the cart microservice. Another way to address this challenge is to build aggregate microservices that incorporate business capabilities. This requires coordination between different entity microservices to address a variety of business use cases, as shown in the following diagram:

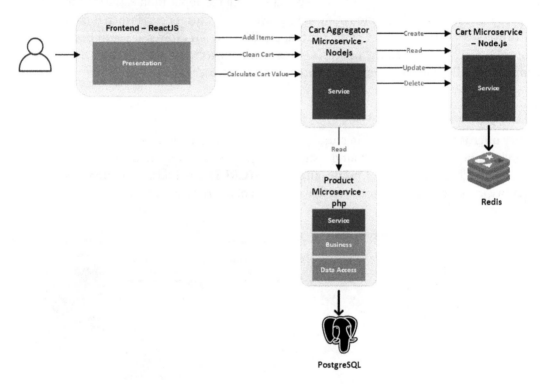

Figure 3.3 – Aggregator microservice

In the preceding diagram, the cart aggregator microservice has been introduced to incorporate additional business capabilities. Let's take a look at this in more detail:

- **Add Items** is a business capability that allows customers to add multiple items to the cart. You can enhance this functionality by checking the inventory before placing these items in the cart.

- **Clean Cart** is a business capability that's exposed to the end user to clear the shopping basket rather than individually remove items from the cart.

- **Calculate Cart Value** is a business capability that helps customers view the total value of the shopping basket. The cart microservice invokes the methods that are available in the product pricing microservice to fetch product pricing. This microservice can then be enhanced by adding features such as **CalculateShippingCost**.

The following diagram depicts a microservice architecture that involves entity microservices and aggregator microservices:

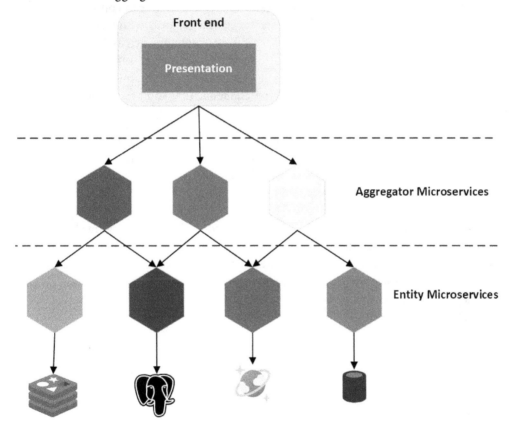

Figure 3.4 – The entity microservice and aggregator microservice in action

The preceding diagram demonstrates the interaction between multiple entity microservices and aggregator microservices in an application. As entity microservices only offer limited functionality, most of the features are implemented in aggregator microservices, which results in creating dependencies between multiple microservices. It also introduces semantic coupling, where a change in one microservice introduces changes in other microservices, which then adds communication overhead that impacts the microservice's overall agility.

The aggregator microservice

In a microservice architecture, there are different ways that a client can communicate with different microservices. In a smaller project, you can get away with directly invoking microservices, specifically when you have a server-side client app such as ASP.NET. However, dealing with clients such as desktop apps and mobile apps is more difficult as each client may have its own limitations and requirements. Maintaining multiple connections with these microservices is another challenge that you need to be aware of as you start adding more microservices.

As the application becomes more complex, it's not uncommon to see microservices breaking the rules of the bounded context and adding extra functionality that doesn't belong to the microservice. Handling cross-cutting concerns such as data transformation, authentication, authorization, and protocol translation should not be part of a business microservice; rather, it should be delegated to a separate microservice. We will call this an **API gateway microservice**. This microservice is responsible for managing client connections and load distribution across different microservices, along with handling some cross-cutting concerns. This approach helps us reduce the number of invocations from the client to address parts of the chatty behavior of the architecture. The following diagram depicts the direct interaction between the API gateway microservice and other microservices in the system:

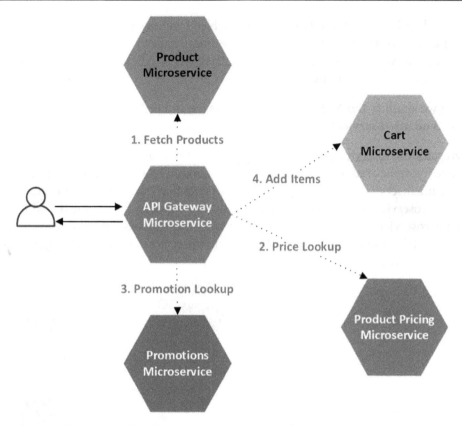

Figure 3.5 – An API gateway microservice with cross-service queries

In the preceding diagram, the API gateway microservice invokes different microservices before adding items to the shopping cart. The sequence of calls is as follows:

1. The API gateway microservice invokes the **product microservice** to fetch the product information specified by the customer while browsing the website.

2. The API gateway microservice invokes the **product pricing microservice** to fetch pricing information about the products that have been specified. If you are selling products in different countries, then the API gateway microservice may have to perform currency conversion as well.

3. The API gateway microservice invokes the **promotion microservice** to fetch all the applicable promotions for customers and apply discounts before compiling the final pricing.

4. Once all the necessary information has been gathered from different microservices, the API gateway microservice will add items, along with their prices and promotions, to the shopping cart.

In one way, the API gateway microservice has enabled business microservices to focus on their intended purpose rather than making synchronous calls to other microservices, but it has also increased the responsibility and complexity of the API gateway microservice. This raises two important questions:

- **Can we simplify the API gateway microservice and decouple it from backend microservices?**

The aggregator microservice pattern can help you build microservices that are responsible for orchestrating communication between different microservices. The aggregator microservice acts as a single point of contact that's responsible for calling different microservices and gathering data to perform a business task, while the API gateway microservice is dedicated to addressing cross-cutting concerns. The following diagram shows the interaction between different microservices, including the aggregator microservice and the API gateway microservice:

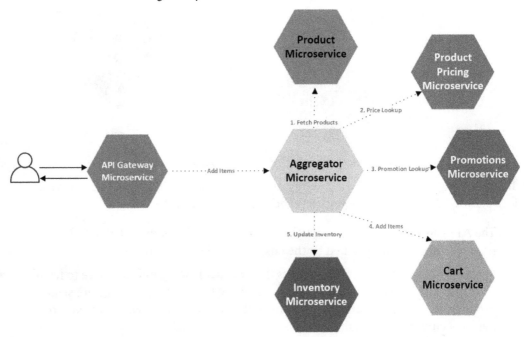

Figure 3.6 – The role of the API gateway and aggregator microservice

In the preceding diagram, the API gateway microservice is acting as a gateway for the backend microservices, allowing different clients to connect and consume microservice functionality. The client is unaware of any of the microservices, or the architecture being used to build these microservices. The API gateway microservice is also responsible for addressing cross-cutting concerns. The aggregator microservice can act as a façade for the API gateway microservice. The aggregator microservice invokes the necessary microservices to fetch the relevant information before populating the shopping cart.

Building an API gateway microservice is not recommended. Instead, you should use an API gateway as a service that can handle scalability and availability concerns for you. Azure API Management is a managed service that can help you publish your APIs to different clients for consumption. Also, it is recommended that you have multiple API gateways exposing limited business capabilities or addressing a particular client type (web, mobile, or tablet). Multiple API gateways will be covered in detail in *Chapter 6, Communication Pitfalls and Prevention*.

- **Can we address the chatty nature of the architecture?**

The materialized view pattern helps you keep a local copy (read model) of the data of other microservices that's necessary for the primary microservice to be more self-sufficient. This pattern helps in eliminating calls to other microservices to increase the response time and reliability of microservices. Adopting this pattern will result in duplicating data across microservices, though duplicating data in cloud-native applications is not considered an anti-pattern. Make sure that only one microservice owns the data; the other microservices can have read models of the data. The read models are synced between microservices. Read models are usually synced using asynchronous messaging patterns using publish/subscribe patterns. The read models are designed to offer better performance using additional indexes, aggregates, and data storage formats to reduce the processing time. The materialized view needs to be published for the first time to make sure that the necessary data is available for the primary service to function. One important aspect that needs to be considered while building a materialized view is the cost of keeping multiple copies of data and the business impact if the data in a materialized view is stale.

The following diagram demonstrates the interaction between the cart microservice and the inventory microservice when using the materialized view pattern:

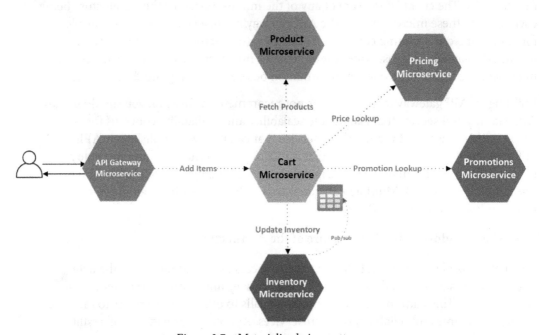

Figure 3.7 – Materialized view pattern

In the preceding diagram, the cart microservice keeps a read model of the inventory microservice, where it uses the pub/sub model to receive updates from the inventory microservice. Once the items have been added to the shopping cart, the cart microservice notifies the inventory microservice to reduce its count, which informs all its subscribers. You can see that other microservices are still accessed via the same communication pattern, which is due to the dynamic nature of these microservices (sub-domains). For example, pricing and promotions are more dynamic, and changes to pricing or promotions may have business implications. In such cases, it's a good idea to prefer fresh data over materialized views. You should implement eventual consistency to sync the read model, and also consider implementing a compensating transaction pattern to address any data consistency failures. Let's have a look at another use case, where materialized views are more useful:

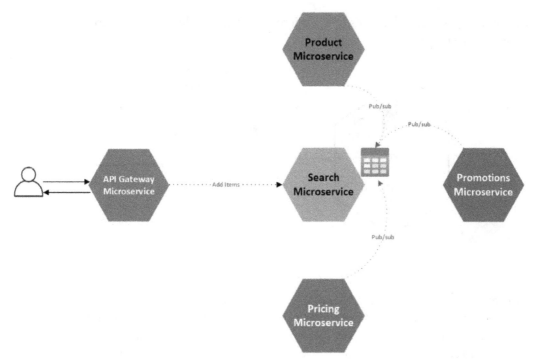

Figure 3.8 – Materialized view of the search microservice

In the preceding diagram, the search service is using a read model to store data from different services to become self-sufficient while serving its users. Also, it's important to understand that presenting stale data in search results may not have the same implications as the cart microservice.

Nanoservices

Nanoservices is an architecture pattern where services are designed to be more fine-grained compared to microservices. These services are responsible for doing strictly one task. Nanoservices are built as LEGO® blocks that can be used alongside other nanoservices to deliver functionality.

Nanoservices have the following characteristics:

- Small and fine-grained
- Independently deployable
- Testable
- Reusable

The following diagram illustrates how a microservice can be decomposed into various nanoservices:

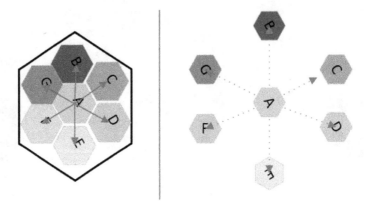

Figure 3.9 – Nanoservices

In the preceding diagram, a microservice is being decomposed into multiple nanoservices to achieve finer granularity, where nanoservice **A** can communicate with the other nanoservices in the ecosystem. This communication pattern is not limited as any of these nanoservices can invoke other nanoservices in different ways to enable parallel processing, pipelines, workflows, and events. Each nanoservice can be built, tested, and deployed independently. In nanoservices, a nanoservice can be reused by other nanoservices to perform a function.

A nanoservice is too fine-grained compared to a microservice. A nanoservice has a single responsibility, but it doesn't follow the principles of domain-driven design and bounded context. A business capability is usually scattered across different nanoservices, which encapsulates fragmented business logic that results in communication overhead between nanoservices. This affects the ability of one nanoservice to deliver any meaningful functionality. It also affects engineering velocity and increases maintenance overhead.

Teams should focus on building microservices by isolating business capabilities rather than greedily approaching every service to become a microservice, resulting in a huge number of nanoservices. Understanding *domain-driven design* is the key to building services with a bounded context to deliver a business capability.

> **Note**
>
> Nanoservices have their space in the software world and in my opinion, they have yet to see their prime. Serverless technology is playing an important role in making nanoservices mainstream and driving their adoption. These days, more and more architectures are being designed to include microservices and nanoservices.

When designing microservices, it's important to understand the business domains that offer business capabilities. Services that are too fine-grained that offer limited functionality often introduce dependencies that neutralize the benefits of microservices and give birth to distributed monolithic architectures. In the next section, we will explore how overusing frameworks and technologies can create maintainability issues for teams, as well as how this can be addressed with a carefully thought-out strategy and a shared governance model.

Overusing frameworks and technologies

Adopting microservices comes in different flavors, where on one hand, teams are learning and incorporating new tools, technologies, and frameworks. On the other hand, teams are building custom tools to facilitate their development and operations. Having many options is good, but not having a shared governance model that can help evaluate tools, technologies, and frameworks is a perfect recipe for failure. This can affect the overall agility of teams, since there is every possibility of shifting their focus from delivering business value to investing time in infrastructure and tooling.

Microservices provide teams with autonomy, where teams can choose the right technologies for their architecture to help them increase engineering velocity. Learning and evaluating different operating environments, programming languages, orchestration platforms, event platforms, and security are important aspects of microservices tooling. These tools, technologies, and frameworks have various capabilities that help organizations build a microservice architecture. As microservices is an architecture style, it doesn't provide any specific guidance on tooling that may result in inconsistent development practices and adding more complexity to the system. For example, the **circuit breaker pattern** is a common pattern that can be implemented as part of the microservice's code base, third-party library (Polly or Netflix Hystrix), or service mesh (Linkerd or Istio).

It's a good practice to have a consistent way of applying these patterns across the code base, which will help reduce code and increase the maintainability of the application. One of the widely used patterns for evaluating tools, technologies, and frameworks is the **capability/feature matrix**. As an example, we have provided an evaluation of different service mesh technologies by using the capability/feature matrix here. The concept of a service mesh will be discussed in *Chapter 6, Communication Pitfalls and Prevention*:

Features	Istio	Linkerd	Consul	Traefik Mesh
Architecture				
Platform	KBs & VMs	Kubernetes	KBs & VMs	Kubernetes
Ingress Controller	Istio Ingress or Istio Gateway	Any	Envoy and Ambassador	Any
High Availability	Yes	Yes	Yes	Yes
Secure Communication				
mTLS	Yes	Yes	Yes	
Certificate Management	Yes	Yes	Yes	
Authentication and Authorization	Yes	Yes	Yes	
mTLS	Yes	Yes	Yes	
Communication Protocals				
TCP	Yes	Yes	Yes	
HTTP/1.x	Yes	Yes	Yes	
HTTP/2	Yes	Yes	Yes	
gRPC	Yes	Yes	Yes	
Traffic Management				
Blue/Green Deployments	Yes	Yes	Yes	
Circuit Breaking	Yes	Yes	Yes	
Fault Injection	Yes	Yes	Yes	
Rate Limiting	Yes	Yes	Yes	
Observability				
Monitoring	Yes (Prometheus)	Yes (Prometheus)	Yes (Prometheus)	
Distributed Tracing	Yes	Some	Yes	
Other Features				
Ops Complexity	High	Low	Medium	
Deployment	Install via Helm and Operator	Helm	Helm	
Multi-cluster	Yes	No	Yes	
Testing	Yes	Limited	No	

Table 3.1 – Comparison of service meshes

In the preceding table, we have illustrated different factors that can help you evaluate different service meshes. This will help you choose the one that's the most appropriate for your project. You should also consider including technology support and community adoption as parameters for evaluation. You can apply a similar technique to evaluate other technologies – for example, evaluating different log management solutions, different NoSQL databases, or caching technologies.

Microservices is an evolutionary architecture, which means that the architecture should evolve as new requirements arise. To begin, teams should concentrate on the fundamentals of microservices to define and decompose them. It is preferable to begin with the tool and technologies that are required and then adopt new technologies as the need arises, after careful consideration. In the next section, we will explore how avoiding abstracting common microservices tasks can affect a team's ability to reduce complexities and adopt change.

Not abstracting common microservice tasks

Abstracting microservices is a practice used in the implementation of microservice architecture. Mostly, the abstraction layer sits on the API gateway layer, which provides various cross-cutting concerns for abstracting common tasks. However, there are various tools available in the market that allow abstraction without adding abstraction to the API gateway layer and help with performing common tasks.

Not abstracting microservices seems to be an anti-pattern since it requires learning effort at the team level, plus it introduces challenges concerning the limitations of the underlying technology and the platform you are targeting. For example, let's say you are building a service in the Go language, and the message broker doesn't support the SDK to communicate with that service. In this case, if we can create an abstraction layer that communicates with the message broker, the service will help the team consume it, instead of reinventing the wheel. If the microservice architecture consists of polyglot technologies, you can imagine the effort that would be required.

Microservices development is challenging as there are limited tools and programming model runtimes for building distributed applications. These programming model runtimes have limited language support and tightly controlled feature sets that target specific infrastructure platforms with limited portability across clouds and the edge. Dapr addresses all these concerns by abstracting the backend services to make developing microservices easy. The following are a few of the benefits offered by Dapr:

- Dapr provides the common patterns for building distributed applications as building blocks that can be reused across the microservice application. These building blocks help in externalizing dependencies and allow developers to focus on business domains by incorporating business capabilities to deliver value to their customers.

- Dapr helps in building cloud-agnostic microservices using any language. This helps simplify the process of developing microservices by keeping applications portable.

- Dapr is an open source and cross-platform framework that's highly extensible and works with any programming language using HTTP and gRPC APIs.

- Dapr also provides a language-specific SDK to make development easy in different languages, rather than just relying on HTTP and gRPC APIs.

- Dapr performs service discovery on behalf of microservices to allow them to call each other without specifying their location. Dapr also performs protocol translation between services and automatically enables communication using mTLS.

Dapr is one of the frameworks that supports abstracting your microservices and offers a set of building blocks to accommodate common microservices needs. We'll explore this in detail in the next section.

Dapr – LEGO® for microservices

Dapr is an open source distributed application runtime that addresses the complexities of distributed systems. It also allows you to dedicate your time to focus on business logic, thus increasing developer productivity. Dapr is a portable and event-driven runtime that facilitates different aspects of building microservice-based applications on the cloud and the edge.

The Dapr runtime can be operated in self-hosted mode or Kubernetes mode. In Kubernetes mode, Dapr injects a sidecar into the pod to enable the sidecar to make service calls on behalf of the application. This makes the application logic simple and lightweight, where Dapr takes care of the overhead associated with the distributed application architecture. Applications are written in different languages, which means they can use the available SDKs to consume the Dapr runtime via HTTP or gRPC APIs. Dapr offers its capabilities in the form of building blocks, each addressing a set of common challenges related to distributed systems, as illustrated in the following Dapr architecture:

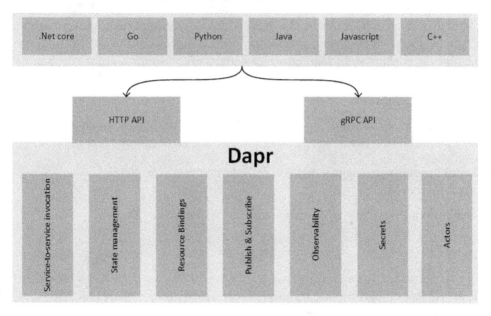

Figure 3.10 – Dapr architecture

Each building block provides a specification (API) which can have multiple implementations, implemented as Dapr components. This helps you choose the right implementation that's tailored to your application. We will discuss each building block in more detail in the following sections.

Service-to-service invocation

In a microservice architecture, different microservices communicate with each other to complete a business task. Service-to-service invocation helps in addressing the common communication challenges regarding service discovery, distributed tracing, error handling, and retries.

State management

Microservices are designed to tolerate failures. These microservices have internal states that need to be preserved beyond the lifetime of individual microservice instances to make these microservices fault-tolerant. Dapr provides the state management building block to allow you to use any key/value store to manage the microservice's state.

Publish and subscribe

The publish/subscribe pattern enables decoupled interaction between microservices, where interested microservices subscribe to individual topics and listen for messages that have been published by different microservices. Dapr provides the *publish/subscribe* building block to enable the publish/subscribe pattern of communication.

Resource bindings

Bindings provide a way of extending Dapr functionality by enabling external systems to communicate with Dapr-enabled applications. There are two different types of bindings:

- **Input bindings** are triggers that enable Dapr microservices to react to external events.
- **Output bindings** allow us to invoke external systems using Dapr.

Actors

The Dapr actor runtime is an implementation of the virtual actor pattern. It's considered an isolated, single unit of compute. These actors receive requests that are executed individually in a single thread. Once the execution is complete, the actor can be garbage-collected based on the configuration.

Observability

Dapr provides an observability building block to help you observe the behavior of Dapr-enabled microservices using metrics, logs, and traces. Dapr also supports distributed tracing by correlating traces across service boundaries to give you complete visibility of service interactions.

Secrets

Dapr's secrets building block provides a consistent set of APIs for integrating a variety of secret stores, including HasiCorp Vault, Azure Key Vault, and Kubernetes secrets, with your microservices. Dapr-enabled microservices can call these APIs to retrieve secrets.

> **Note**
>
> You can learn more about Dapr at `https://dapr.io/`.

Dapr can be hosted in multiple environments, including self-hosted environments for local development or to deploy to a group of VMs, Kubernetes, and edge environments, such as Azure IoT Edge. In the following section, we'll go over the significance of being knowledgeable about microservices platforms.

Lack of knowledge about the microservice platform

A microservice platform is a fully managed platform that orchestrates a cluster of machines to deploy microservices. Kubernetes is one such platform for hosting microservices that can help you achieve better scalability, availability, and resilience for your applications. Managing and securing microservices platforms at scale requires significant investment in skills and tooling. Organizations have formed dedicated teams to manage multiple clusters in their production environment.

In the next subsections, we will learn how the adoption of containers, cloud-native applications, and micro platforms can accelerate the development and operations of microservices.

Embracing containers

Containerization is the process of packaging applications with their dependencies (binaries, libraries, and configuration files), excluding the operating system, which makes them very light and portable. The package is known as a container image and is stored in the container image registry. A container image can be deployed across multiple environments (private cloud, public cloud, local machine, and multi-cloud) consistently while retaining its functionality to run multiple container instances to serve its users. Each container instance runs in its isolated user space to make sure that the failure of one container does not affect other containers in the ecosystem.

Virtualization has certainly helped organizations with the agility to allow them to create multiple environments and make use of unused capacity available on the hypervisor host. However, hosting applications on virtual machines is still a lengthy process that includes installing guest operating systems, libraries, and applications. It is a recommended practice to dedicate hosts for individual applications to avoid any dependency hell. The following diagram shows a comparison between hosting applications on virtual machines and containers:

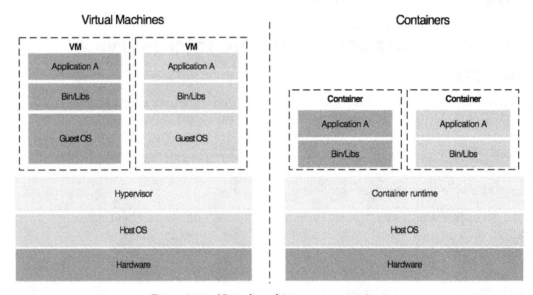

Figure 3.11 – Virtual machines versus containers

Compared to virtual machines, containers are very small in size and share the same OS kernel. Multiple containers can be hosted on a single virtual machine to increase their density. In most cases, containers are disposable, where new changes can easily be rolled out by replacing old containers with new containers, without this affecting the underlying operating system. There is one exception, however, and that is when these containers run stateful services such as databases or persistent storage. With the widespread use of containers across businesses, legacy apps that were originally developed as stateful applications are being refactored to operate as containers. Converting such legacy apps into stateless containers will need a significant upfront investment. Due to these factors, container vendors consider stateful containers as an important aspect of the container platform.

Containers have enabled organizations to simplify the application life cycle for the delivery of their software. Due to containerization, organizations are embracing continuous delivery to release software with speed and security. However, running containers in production requires investment in tooling to automate container management at scale to reduce the operational workload. Container orchestrators help you achieve better availability, scalability, resilience, and resource management for your container workloads. Kubernetes, Docker Swarm, Service Fabric, and Mesos are a few such container orchestrators.

> **Note**
> Recently, Kubernetes has become the de facto platform for hosting container workloads. Kubernetes provides various features that you can explore at `https://kubernetes.io/`.

Embracing the cloud-native approach

The *cloud-native approach* is about building applications that support rapid change, large scale, and resilience to maximize the capabilities offered by cloud providers. These cloud providers have a presence in multiple geographic regions to help you establish highly available and resilient environments across different regions.

These cloud providers offer different type of services (IaaS, PaaS, and SaaS), which can help you build a variety of infrastructures for your applications. PaaS is a platform offering that automates most of the administrative and management tasks you need to perform regularly to manage your application infrastructure. The cloud-native approach emphasizes using PaaS services, which offer better management and SLAs compared to on-premises and IaaS environments.

As Kubernetes has become the de facto platform for cloud-native applications, many cloud providers have a serverless offering for Kubernetes, which is an enterprise-ready Kubernetes platform that addresses the concerns around continuous integration, continuous delivery, security, and governance. **Azure Kubernetes Service (AKS)** and **Azure Red Hat OpenShift (ARO)** are the managed services offered by Azure. Azure also offers **Function-as-a-Service (FaaS)** through Azure Functions for hosting smaller applications that are available on demand.

Many enterprises are realizing the importance of hybrid environments for hosting their workloads across a variety of environments. This can help them address scalability, regulatory, and compliance concerns for their customers. In the cloud-native world, enterprises are continuously exploring ways of managing hybrid cloud estate easily and consistently. *Azure Arc for Kubernetes* enables any *CNCF* certified Kubernetes cluster to be managed using a single pane of glass. Now, organizations can deploy their cloud-native workloads on their choice of cloud provider and manage their workloads with ease using Azure Arc.

The cloud-native approach also emphasizes adopting modern application design principles for building containerized microservices with asynchronous messaging and modern communication protocols (gRPC and Thrift). Another important aspect of the cloud-native approach is to enable observability across different microservices that are deployed across a cluster of nodes. Monitoring platforms help you centralize logs and telemetry for monitoring and alerts. Automating code and infrastructure is a big part of the cloud-native approach when it comes to building an immutable infrastructure that can be deployed across environments with consistency. Continuous integration, continuous delivery, and **Infrastructure as Code (IaC)** are the widely adopted practices in the cloud-native world.

Embracing micro platforms

The *micro platform* can be seen as a small microservices platform that is much easier to create, manage, and dispose of compared to the microservice platform. Micro platforms are considered as smaller landscapes that only target a set of services that are part of one application. Many organizations have adopted the practice of managing multiple microservices platforms, where each application has its own micro platform instead of having one large platform where all the services are provisioned. One of the advantages of a micro platform is that it can protect services from cascading failures.

These platforms are created using IaC practices so that they remain identical and reproducible. In some cases, you can deploy these platforms on your local machine. The micro platform helps in increasing the autonomy of teams by empowering them to manage the platform, along with microservices, and move faster as they no longer have to rely on other teams to provision their microservices. The micro platform helps organizations be more resilient compared to a single microservices platform, where they have the flexibility to provision multiple micro platforms in hybrid environments or even across different cloud providers to achieve better resilience. The following diagram shows a comparison between the microservice platform and micro platforms:

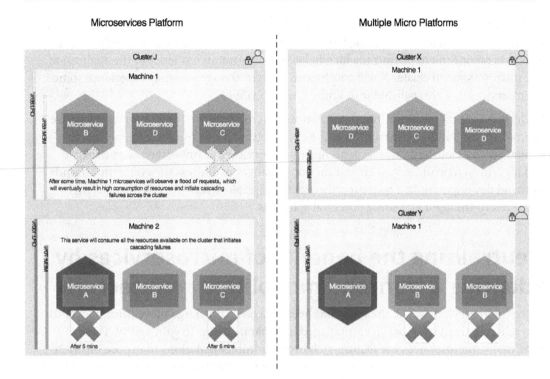

Figure 3.12 – The microservice platform versus multiple micro platforms

In the preceding diagram, the microservice platform is managed as a single cluster of machines that are managed by a dedicated team, responsible for maintaining and operationalizing the cluster. In a microservice platform, microservices are deployed across the cluster, thus sharing a common space for consuming resources that can inadvertently affect other microservices.

In the aforementioned scenario, *Microservice B* has a bug that results in consuming more CPU and memory, which eventually causes failures for other microservices on *Machine 2*. Since *Machine 2* is no longer available to serve user requests, all the user requests are redirected to *Machine 1*, which will result in more resource consumption on *Machine 1*. At the same time, the microservice platform will be trying to host the required number of microservices on *Machine 1*; that is, *Microservices A*, *B*, and *C*.

In the second scenario, we have two micro platforms hosting a different set of microservices, which is more resilient to cascading failures. Failure of one microservice can only affect the resources on one micro platform, which helps with the overall resiliency of the application. The microservices in cluster X will continue to operate, though you may experience some microservices that are available with limited functionality.

Understanding containerization, orchestrators, micro platforms, and other cloud-native approaches for building microservices is key for setting the foundation for success in a microservices journey. The frontend of a microservice is as important as its backend, and it's critical to explore how cross-functional teams can address the concerns around frontend development in a microservice architecture. Now, let's discuss the challenges of the micro frontend and some potential solutions.

Neutralizing the benefits of microservices by adopting a frontend monolithic architecture

In the microservices world, the focus has been on converting large monolithic backends into self-contained microservices, managed by separate teams to gain agility. These days, many microservices use feature-rich browser implementations to expose business functionalities to their end users. Mostly, there is a separate team that's responsible for building application user interfaces by consuming APIs exposed by microservices. Over time, these frontends grow in complexity, resulting in a frontend monolithic UI. Hence, the benefits of microservices are somehow constrained due to the architectural style that's followed at the start. To overcome this challenge, you should introduce a *plug-and-play architecture*, where each team can build their own frontends. *Figure 3.1* depicts a monolithic frontend that interacts with multiple microservices. To incorporate multiple microservices, frontend teams need to work with multiple backend microservice teams, which increases the communication overhead for managing different parts of the UI that are affecting its agility and productivity. The real challenge is to design an architecture that supports decoupling, scalability, and independent deployment.

The micro frontend addresses the drawbacks of a monolithic frontend, which allows teams to reduce communication overhead, increase autonomy, and help organizations scale. We'll learn about micro frontends in the next section.

Micro frontends

Using a micro frontend is an architectural style that focuses on scaling the development of frontends by bringing complete autonomy to a business domain. This enables teams to deliver independent UIs that can be combined to build a product. In essence, each team is responsible for delivering their UI components, pages, or JS files. The main application is built as a container of UI components, where each component can be developed using different technologies. The following diagram is a representation of a micro frontend architecture:

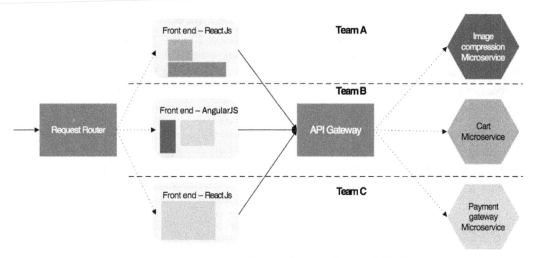

Figure 3.13– Microservice platform with micro frontends for the web

In the preceding diagram, each team is responsible for developing UI components that belong to their subdomain. Pages are composed using different components produced by different teams, where the components can communicate with each other using predefined interfaces, and routing is managed either on the client side or the server side. Single-SPA and web components are the frameworks that can facilitate the development and composition of multiple UI components as a single product. Also, the adoption of micro frontends enables cross-functional teams to own the complete business domain, where the microservices, API gateways, and micro frontends are owned by one team.

Summary

In this chapter, we learned about various architecture pitfalls while building microservices. We discussed the layered architecture, its complexities, its benefits, and how it can be applied to microservices. We also discussed how entity microservices and nanoservices can lead to complexities in your microservice architecture, all of which can be addressed by implementing different strategies, such as API gateways and aggregator patterns. Which tools, technologies, and frameworks you select should be well-thought-out and discussed as part of the shared governance model across the teams to bring consensus.

Then, we discussed how we can address the architecture intricacies of microservices using Dapr. After, we discussed how having a lack of knowledge about container technologies, orchestrators, and cloud-native platforms can slow down the adoption of microservices and how we can accelerate this adoption by embracing them. Finally, we discussed the drawbacks of monolithic frontends and how they can neutralize the benefits of the microservice architecture, which can be addressed using micro frontends.

In the next chapter, we will discuss the common pitfalls while migrating monolithic applications to microservices and how we can address the different aspects of refactoring brownfield applications. We will also discuss the role of availability, scalability, and reliability while building microservices.

Questions

1. Explain the aggregator microservice pattern.
2. How can you evaluate different tools, technologies, and frameworks for microservices?
3. What are the main building blocks of Dapr and how can they help when you're building microservices?

Further reading

For more information regarding what was covered in this chapter, take a look at the following resources:

- *Practical Microservices with Dapr and .NET*, Davide Bedin
- *Cloud-Native with Kubernetes*, Alexander Raul
- *Developing Microservice Architecture on Microsoft Azure with Open Source Technologies*, Arvind Chandaka and Ovais Mehboob Ahmed Khan
- *Micro Frontends in Action*, Michael Geers

4
Keeping the Replatforming Brownfield Applications Trivial

The microservices architecture has been adopted at a rapid pace in industry recently. Many organizations have been adopting a microservices architecture because of its advantages and exploiting it by designing new applications (greenfield applications) or turning existing monolithic applications (brownfield applications) into microservices architectures. Replatforming monolithic architectures requires that you have a detailed comprehension of microservices. You will shift your attitude regarding how to use these practices to help with your journey.

In this chapter, we will not only cover the pitfalls of replatforming brownfield applications but also learn how to decompose a monolithic application into a microservices architecture, the importance of having availability, reliability, and scalability in your applications, and how to build self-contained services.

The following topics will be covered in this chapter:

- Not knowing the fundamentals when replatforming brownfield applications
- Overlooking availability, reliability, and scalability when designing the microservices architecture
- Sharing dependencies using older techniques

This chapter aims to help you understand the process of moving from a monolithic application to a microservices architecture, as well as the pitfalls to be aware of and avoid along the way. In addition, we will also illustrate several key trends to consider when designing a microservices architecture.

Not knowing the fundamentals when replatforming brownfield applications

When it comes to translating a monolithic application into a microservices architecture, ignorance is not bliss. Not all microservice projects start as greenfield applications, where the architects and developers have a wide set of options to design the architecture and build the system without refactoring or replatforming the whole monolith. Monolithic architectures have been common in the industry, but there are still certain drawbacks if the same techniques are adopted when building a microservices architecture.

In the following sections, we will discuss the significance of the microservice architecture, the factors to consider, and the decoupling approach. Let's get started!

Why we should go for a microservice architecture

Let's look at some of the things we should consider when transforming monolithic applications into a microservice architecture.

Easy to develop and test but you must restrain developers from adopting new technologies

Monolithic applications are easy to develop and test because they are likely based on a set of related or well-tested technology stacks, and they have a shallower learning curve in terms of technology adoption. Since the projects are directly referenced, there is no network latency and they run as a single process. Usually, monolithic applications have a single database that contains all the database objects, such as tables, stored procedures, and the functions that are used to perform **Create, Read, Update, and Delete (CRUD)** operations. This is one of the benefits of having everything in one place, and it's good for rapid development for smaller companies. However, as the company grows, monoliths slow the pace of innovation and development as developers end up investing much more time in trying to understand and handle scenarios, such as where members of other teams are tinkering with data that they own, more operational work because of a huge blast radius from minor issues, and so on. Another drawback is that it limits developers in bringing in new technologies when building new features; they must stick to the same technology the application is based upon. As the application grows, it becomes more monolithic, maintenance becomes difficult, and adaptability decreases. In the long run, it not only discourages developers from using emerging technologies but also becomes difficult to update, modify, release, or deploy.

The n-tier architecture provides SoC but has tight coupling

The n-tier architecture is one of the most widely adopted architectures that separates concerns between layers. Its tight coupling, on the other hand, makes it more difficult to release faster to the market and requires rigorous regression testing before release. Changes in one place may have a cascading impact on other components, and altering entails a large amount of risk.

A monolithic application is usually based on the n-tier architecture, where the presentation layer contains the frontend components such as pages, the view model, CSS, and JavaScript files. The presentation layer communicates with the service layer or sometimes directly with the business layer by adding a direct reference to the business layer. The business layer contains the core application logic and communicates with the data access layer to perform database operations.

The following diagram shows the difference between the monolithic architecture and a microservice-based architecture. The monolithic architecture shows two kinds of n-tier architectures, where the left-hand side shows a direct reference to the business layer inside a presentation layer, which enables direct dependencies between layers. On the other hand, the right-hand side of the monolithic architecture shows loose coupling by having a service layer exposing APIs to allow the presentation layer to communicate with the service layer over HTTP/HTTPS, HTTP/2, or gRPC protocols:

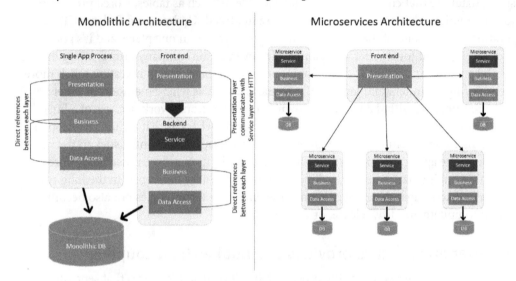

Figure 4.1 – Comparing the monolithic and microservices architectures

In a microservices architecture, the system is decomposed into a set of various fine-grained services, where each service is a RESTful API exposing an endpoint over HTTP. The presentation layer contains various frontend components and communicates with the backend services over standard protocols, as mentioned previously.

Simple deployment but expensive scaling

Monolithic applications are simple to deploy since the application is packaged into a single assembly or artifact that can be deployed. Unlike microservices, where the application is decomposed into several services, manual deployment is not an option. The following diagram shows the difference between the monolith architecture and the microservices architecture for deployment:

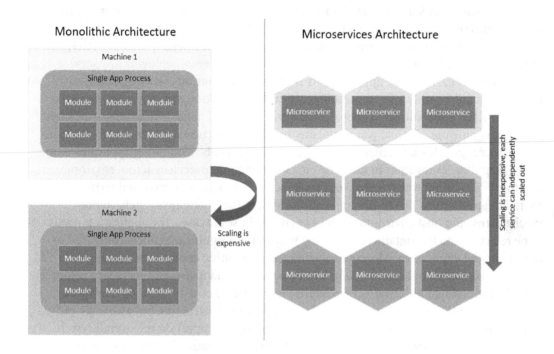

Figure 4.2 – Comparing the monolithic and microservices architectures regarding scalability and deployment

The preceding diagram shows a comparison between deploying monolithic and microservice-based applications. With monolithic applications, the deployment involves a risk since it has to be deployed as a whole, and if the application is large, changing something in one place may have a huge risk if it is not tested properly.

On the microservice side, manual deployment is not recommended as there are many services, and the manual deployment needs more effort and time. To overcome this challenge, DevOps **continuous integration** (CI) and **continuous deployment** (CD) practices can be used to automate builds and releases. However, deploying any service has a minimal risk since the application is segregated into various services, and if any deployment fails, it only affects a certain part of the application and can easily be rolled back to the last good known version. Moreover, if a manual patch is necessary, you should use the CD pipeline, just like you would any regular production deployment, to not cause any further regressions. Furthermore, containerization enables bundling applications and their dependencies together to facilitate containerized microservices across environments and to enable scaling with low risk and faster scaling. It is difficult to containerize monolith applications due to the application's size, nature, complexity, persistence mechanisms, and host environment dependencies. In some cases, the application needs to be refactored to be containerized, which may require significant efforts compared to rebuilding the application. Moreover, scaling the monolithic application running on a VM requires a capacity similar to that of a VM to scale out, and it also requires more resources regarding individual microservices, resulting in this being cost-prohibitive. However, with microservices, we can not only use the container infrastructure but also scale out individual services independently, without this having an impact on the overall system and the overall business functionality. Another huge drawback of monoliths is that it's super hard to understand the CPU and memory profile, as well as optimize the workload, since so many modules are running at the same time. With microservices, that problem vanishes.

Having understood the motives behind replatforming, let's move on and discuss the factors we must keep in mind when performing replatforming.

Factors to consider when replatforming monolithic to a microservices architecture

The process of replatforming a monolithic application is not straightforward. To understand this, let's look at some of the aspects we should consider when transitioning from a monolithic architecture to a microservices architecture.

Knowledge of business domains

Understanding the application's business domain is the core factor to consider when replatforming. Architects and developers should spend proper time not only understanding the application's underlying architecture and design, but also on how the data can be decomposed based on business domains. Sometimes, the UI screens are useful for understanding the service boundaries.

Focusing only on technology when decomposing microservices and working in an isolated way without involving the business stakeholder, who knows the business and can guide the development team to decompose them based on business capabilities and the domain, is not a good practice. Giving less value to this space would eventually affect the design of the overall architecture and make it fail to achieve agility. If the service is not decomposed properly, it creates dependencies, so changing a particular function requires multiple services to be updated and becomes a kind of distributed monolith. You must be aware of the change in infrastructure.

Microservice-based applications are decomposed into several services. Each service runs independently, without having an effect on the other service. This is what brings agility and reduces the risk of changes. However, compared to monoliths, they are simple to host and manage. With microservices, we need to upgrade the infrastructure to support deploying each service into its own separate space. We can leverage a container infrastructure that provides isolation and easy scaling for each service, where each service can run inside its own container and share the same host operating system kernel and system resources. You can also orchestrate the containers using managed Azure Kubernetes Services to leverage high availability, auto-scaling, and increased reliability.

Another factor to think about is network latency, which can have a direct effect on an application's efficiency if the system or the underlying infrastructure isn't up to par. Each service has its own database, which is hosted in the cloud or on-premises, and certain transactions can span over multiple services, with some patterns and technologies being used to create dependable services and utilize asynchronous communication throughout services. Before beginning this transition, it is vital to raise awareness about the infrastructure improvements that are needed.

Choosing the right technology and embracing the learning curve

With microservices, one of the benefits is the adoption of technology. You can use any technology to build any service, unlike monolithic applications, where you have to rely on the underlying technology, regardless of if it is deprecated or obsolete. However, choosing the right technology is important.

When choosing a technology, make sure it supports some libraries to build the service and handle cross-cutting concerns such as implementing logging and instrumentation, data connectors to connect with databases or other storage accounts, and so on. Performing analysis before picking a technology saves you the time you would have spent on reworking and the effort you have put in to learn that technology. Based on all these considerations, always check to see if there is enough community involvement and a forum where you can report issues. For instance, in many cases, open source software companies or individuals maintain their own GitHub sites where the community can contribute, check active discussions, open issues, and report new issues as well. For closed source software, make sure they have a support site where the issues can be reported.

With a microservices architecture, you have the freedom to experiment with emerging technologies and build your services around them. This way, you can not only match your technical competence with current market trends and use the desired technology to develop that service, but if there are some technology-specific limitations, the risk will be lower since it only impacts the business capability offered by that microservice.

Go cloud native

Cloud services are the foundation of cloud-native applications. Cloud-native services provide a better SLA in terms of availability, reliability, and scalability, which is not easy to do with an on-premises architecture and requires huge investment in the infrastructure. With microservices applications, there are numerous services, and scalability and availability are the most important concerns. Any service that fails to respond would have an impact on the business.

There are few elements of cloud-native applications to consider, as follows:

- Application design should be microservices
- APIs provide access to standardized methods
- Services should utilize container infrastructure
- Manual deployment should be avoided and use CI and CD practices
- DevOps practices should be adopted to adhere to agile development
- Centralized logging and instrumentation should be maintained
- Embrace shift-left testing and start testing as early as possible

Failing to follow any of these measures would make administrating, maintaining, and modifying microservices difficult.

Understanding the difference between rearchitecting and refactoring

When it comes to replatforming applications, rearchitecting and refactoring are two sides of the same coin. The focus of rearchitecting is on the high-level architecture, which includes, but is not limited to, how services interact with one another, what communication protocols are used, and what hosting platform or deployment strategy can be adopted. The following are some of the activities that should be considered when rearchitecting any application:

- Redesigning the architecture to develop microservices
- Focusing on capturing the business value
- Replacing the existing architecture components, frameworks, technologies, or platforms to transform into the microservices architecture
- Leveraging modern application computing concepts
- Embracing cloud-native application practices
- Validating the solution architecture design using a tool such as Microsoft Threat Modeling

> **Note**
>
> To learn more about the Microsoft Threat Modeling tool, refer to the following link: `https://www.microsoft.com/en-us/securityengineering/sdl/threatmodeling`.

Refactoring, on the other hand, is primarily concerned with how to enhance the design of the current code. Its goal is to evaluate what design patterns are used, how the code is structured, and how it may be improved. In the context of a microservices architecture, when refactoring a brownfield application, some of the activities that are usually carried out are as follows:

- Code refactoring to align with business domains
- Using OOP principles where applicable
- Using SOLID principles where applicable
- Using design patterns where applicable

- Code optimization by removing dead code
- Analyzing Big O notation and optimizing code where applicable

> **Note**
>
> To learn more about SOLID principles, refer to the following link: `https://docs.microsoft.com/en-us/dotnet/architecture/microservices/microservice-ddd-cqrs-patterns/microservice-application-layer-web-api-design`.

Both rearchitecting and refactoring assist in transforming a brownfield application into a microservice-based architecture, where rearchitecting focuses on the application's design and solution architecture, while refactoring assists with implementing that architecture through code modifications and other related techniques.

Understanding the difference between core microservices and API microservices

Microservices that are consumed by other microservices are known as core microservices, whereas API microservices are consumed by the frontend. The core microservices are based on a reactive pattern, which is an approach that's used to build a system based on the message-driven architecture and supports resiliency, reliability, and elasticity. The architects and developers should have knowledge about reactive systems and build core services using them. The following diagram depicts the core and API microservices interacting over the Event Bus:

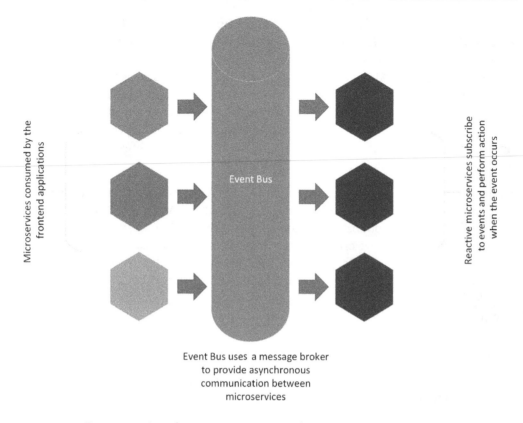

Figure 4.3 – Asynchronous communication between core microservices
and API microservices using an Event Bus

In Azure, various services can be used to build message-driven systems. Some of the names include, but are not limited to, Service Bus and Event Hubs. They both offer a publish/subscribe model where the service can publish messages to a queue, and subscribers can listen to that queue to read the message.

Avoid chatty services

Chatty services contain a long chain of operations and they rely on several services to complete one single transaction. Chatty services not only reduce the agility in adopting change but also heavily rely on the underlying network, database queries, disk reads, and various other factors. Since HTTP communication is stateless, if any service is inaccessible or takes a long time to respond, timeout exceptions may occur. If your services are heavily dependent on each other and communicate synchronously, it becomes a distributed monolithic architecture, and you undergo the same pain when a target service is down. Thus, all the services that are used within the transaction have to be operational at the same time a transaction occurs.

To overcome this challenge, the approach is to decompose services while following **domain-driven design (DDD)** and build services based on business domains. There are scenarios where interaction with other services is still required, which can be solved by implementing a message-driven architecture. There are patterns such as Saga that help you avoid synchronous communication between services and use a message broker to support asynchronous communication. However, this does not mean that synchronous communication is an anti-pattern; it depends on the usage and business scenario. For example, data services, notification services, or workflow services are some examples where synchronous communication could be the right choice.

Versioning also plays an important role when building REST-based services. You should always version a service when the service contract is changed. When versioning, make sure that all changes are both backward and forward compatible. Once all the callers have moved to the new version, we can potentially remove some of the backward compatibility code. This avoids breaking other services that are dependent on that service and gives the developers a chance to update it at their own pace. Furthermore, this also allows you to slowly deploy the new service and enable easy rollback in case of any issues during one box/canary testing.

Development, delivery, and operations readiness

Building a microservices architecture is not simple and requires good knowledge of the technologies and patterns to address challenges. Complexities can emerge in development, delivery, and operations.

Development complexity arises when developers are asked to use different patterns to address challenges. For example, implementing distributed design and service interaction patterns such as Saga or CQRS requires having a practical understanding and knowledge about how to use them and their approaches. Since these patterns are not normally used with monolithic architectures, developers are usually not familiar with them. Furthermore, since any service can be built on any technology that is best suited to the underlying business scenario, it may require a learning curve and adequate preparation to develop the service.

On the database side, since each service can keep its data in its own database, you can easily build a polyglot persistent architecture and use the technology that best suits that business functionality. However, it might take some effort to build the database model. Therefore, to build one service, a developer should know not only the design patterns but also when to use which technology. Considering an e-commerce system, we can choose Azure Table Storage to store cart information instead of using some relational database. However, for order management, SQL Server or any NoSQL databases can be used as well.

Delivery complexity arises when there are multiple services based on different technologies, and each service requires a different set of tasks to build and deploy applications. Container infrastructure solves this problem to some extent since you can containerize services once and deploy them to any environment using DevOps CI/CD practices. Microsoft Azure DevOps provides end-to-end application life cycle management capabilities and offers lots of connectors to build CI/CD pipelines for any technology or platform and deploy to any environment.

> **Note**
>
> You can set up a free account for Azure DevOps Services at `https://dev.azure.com`.

Operational complexity comes when you deploy your services to production or staging environments. Services should be resilient and scalable, and choosing the right platform is very important. If your services are running inside containers, you can use managed **Azure Kubernetes Services (AKS)** to set up a desired state for containers, and also set various metrics to auto-scale your AKS cluster or pods when it suffices for that metric condition. For monitoring, tools such as Azure Application Insights, Grafana, Prometheus, and Jaeger help with monitoring your services and spinning up alerts to send notifications. Organizations should have a shared governance practice in place where they can spend resources to provide company-wide standardization solutions for things such as building, deploying, orchestrating, and monitoring microservices.

Now that you have a thorough understanding of the factors to remember when refactoring monolithic applications into a microservices architecture, let's look at the decoupling approach.

The decoupling approach

The decoupling approach is a process that can be followed when moving from a monolithic application to a microservices architecture, and it undergoes various phases. The first and most important phase is to understand the overall business architecture of an application and to identify the domains. Following DDD practices, you can categorize business domains into core, supporting, and generic domains, and start with the simplest service, which can easily be extracted and have less impact on the overall system. On the other hand, if you already have a new requirement, you can simply build that new requirement as a separate service and integrate it with the existing system. Services that are classified as generic domains are simpler to decouple than ones that are part of the core domain. For example, the authentication service can be categorized as a generic domain since it is not the main business of the application, even though it supports the application by providing user authentication and authorization. Such kinds of services are easy to decouple and the transition is smooth. Going forward, once all the generic services have been decoupled, the next phase you can start is building the supporting domain and then the core domain. The following diagram shows how to decompose the authentication module from a monolithic application to a new microservice:

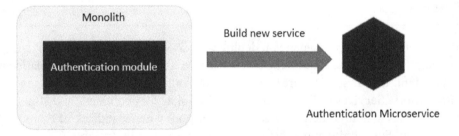

Figure 4.4 – Decomposing an authentication service that is the least dependent on a monolith application

When decoupling services, keep the dependencies as minimal as possible. For example, if the decomposition is done for a particular business domain, take the tables and recreate them in a separate database. You can not only alter the database schema but also use any other persistence technology to keep its domain-specific information. If some transactions span multiple services, start decomposing the service that has minimal dependency on other services.

Since the monolith to microservices transition is not a one-step process, to keep the actual application up and running, you need to replicate the data to let the system work. This is needed until all the services are decomposed as separate services.

The following diagram shows the purchase workflow of a standard e-commerce system. Once the user hits the checkout service, the user is taken to a payment gateway to pay, the order is generated, and an email notification is sent:

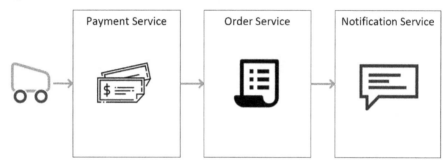

Figure 4.5 – Series of operations that occur when the checkout operation is used in an e-commerce system

Considering the previous example, the first service to transition is a notification service that has no dependency on other services. However, order generation is dependent on this service because when the order is generated, the notification needs to be sent out.

When decomposing services, some services may need to have a call go back to the existing monolithic application to keep the functionality running. This is because if there are other modules of the monolith using the same functionality somewhere in the database or at the application level, we need to build an anti-corruption layer just to synchronize the data and keep the functionality of that monolith working. When building the anti-corruption layer, you can build a new API for the monolith that communicates with the monolith application, and the new microservice communicates with that monolith API instead of having references or code to communicate to the existing monolithic directly. This practice helps isolate the monolithic concepts. The following diagram shows the approach of decomposing the invoice generation service and using an anti-corruption layer to keep it functional:

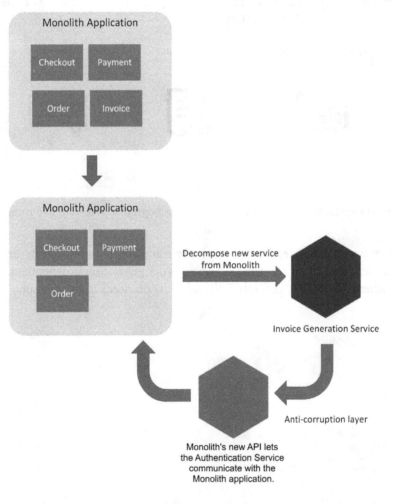

Figure 4.6 – Adding an anti-corruption layer when decomposing the service

Decoupling can be done by vertically splitting the n-tier architecture based on the business domains, decomposing each service one at a time, and then creating a new service for each vertical layer that contains domain-specific models. You must create a separate database and use the right persistence technology to keep its domain-specific information. Once the service has been built, you must modify the frontend so that it consumes the new service's endpoint. Until all the services are decomposed, you can replicate the data to keep the monolith working and implement an anti-corruption layer, also known as a **glue code pattern**. The frontend makes a call to the new service, where the new service makes a call to the new monolith API to keep the monolithic database updated. While decomposing, you can adopt different techniques and patterns to implement distributed transactions, synchronous and asynchronous communications, and so on. Once all the services have been decomposed, the monolithic database can be archived, and each service will have its own persistence store where it can store its domain-specific information. The following diagram depicts a few of the modules of an e-commerce application, as well as the steps to decompose it from a monolithic application to a microservices architecture:

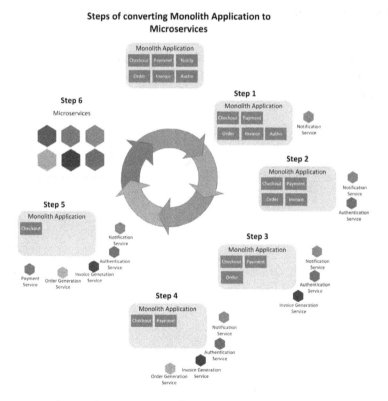

Figure 4.7 – The step-by-step process of decomposing a sample monolith application into a microservices architecture

When decoupling services, you can either reuse some part of the existing monolithic application code in your microservices or rewrite it completely. If the application is tightly coupled with other modules, it cannot be easily decoupled and it is a better trade-off to rewrite it. Every application has its own complexities, and this can be determined based on the efforts needed to decouple them. Also, with monolith applications, it's not necessary for the service you are trying to decompose to be extracted and mapped to the monolith business module.

When decomposing the monolithic application, you must decouple the services that are simpler to decompose and have fewer dependencies. In the preceding diagram, we begin by decomposing the notification service, which is less dependent on other services and can easily be unplugged. Decoupling the most basic service first minimizes the chance of delivery teams failing to correctly operate the microservice and understanding how the decomposition will work. Once the notification service has been decomposed, the authentication service can be decomposed, then invoice generation, and so on. Until all the services have been decomposed, you need to implement the glue code pattern, as discussed previously, to replicate the data to the monolithic database to keep the system functional. Once all the services have been decomposed, the glue code pattern can be removed.

When decomposing, it is critical to understand certain patterns that can be used to build reliable services. In the following section, we'll go over the significance of availability, reliability, and scalability.

Overlooking availability, reliability, and scalability when designing a microservices architecture

As microservices architecture is distributed in nature and involves lots of services communicating over a network, reliability and availability, along with scalability, become important factors. Developers sometimes overlook the benefits of these factors and don't think outside the box to build an architecture that supports these concerns. In a microservices architecture, services have dependencies on other services and along with asynchronous communication, there is also synchronous communication, which requires the other service to be responsive at the time you're performing some operation or serving an end user request. If the service is not available, it affects the transaction. Moreover, when a service is experiencing a lot of traffic, it needs to be scaled out and if the underlying infrastructure does not support quick scaling, this will impact the operation and create inconsistencies across databases. As a side note, you should build systems in a way that they should be capable of handling any minor inconsistencies on their own. For example, if an end user request has to write to two services, a cheaper way than using a transaction is to write to them one by one and handle inconsistencies on the read path; that way, you don't need to bother with transactions as rolling back is very tricky in such scenarios.

Let's take a closer look at why availability, reliability, and scalability are major determinants to consider.

Availability

Availability is the percentage of time a service is operational and handles requests. We need to keep things available to that agree by creating achievable targets that have been pre-agreed or committed to our customers. The availability of a service depends on how frequently the service fails, and how quickly it recovers after a failure. As a general rule, availability can be calculated using the following formula:

*Availability in percentage = (Total elapsed time – Total downtime) / Total elapsed time * 100*

For example, if the total elapsed time is 720 hours and the total downtime of the service is 1 hour, then the SLA will be 99.86%. An organization must agree on the availability of every service with service operators, and if that service is down, they must measure the impact in terms of how many users are affected and how the business is impacted. A few of the most important factors that affect availability are infrastructure, application quality, people, and process/automation. If the infrastructure is good and it provides a better SLA, the application will be available unless there are some exceptions. On the other hand, if the application is thoroughly tested and there are no unhandled exceptions, it will not fail and will continue to run. Unhandled exceptions are one of the most serious issues that affect the application's responsiveness. Furthermore, if people from the organization are not skilled enough to operate the environment in critical times, this may affect the availability of the system during recovery.

When choosing an architecture, it is better to go cloud native. This is because cloud-native solutions offer various managed services that provide better SLA and scalability options. For example, in Azure, there are many managed services, such as Azure App Service, Azure Kubernetes Services, and Azure Service Fabric, that can be leveraged to build and run microservices.

Reliability

Reliability states how well the service will be available in different scenarios. For example, a managed cloud service offers a better SLA, but a buggy program may trigger service outages that are unrelated to the underlying infrastructure. There are few ways to measure reliability, with the most common one being **mean time between failures** (**MTBF**).

Here is the formula for MTBF:

MTBF = (Total elapsed time – Total downtime) / Number of failures

For example, if the total elapsed time was 720 hours, the total downtime was 1 hour, and the number of failures when the service was not responsive or throwing exceptions was 10, then the MTBF will be 71.9 hours.

Apart from MTBF, organizations should also aim to measure the **mean time to repair** (**MTTR**). This is a very important factor for figuring out the recovery time and how the services will respond if any failures occur. Here is the formula for MTTR:

MTTR = Total maintenance time / Number of repairs

For example, if the service failed several times in a month due to transient errors and the total maintenance time spent was 60 minutes, with a total number of 10 repairs, then the MTTR will be 6 minutes.

Various patterns can be used to build reliable services. Some of the most common ones are the retry pattern and the circuit breaker pattern.

The retry pattern

The retry pattern is used to enable resiliency in an application by making multiple attempts to connect to a faulty service that is failing due to transient errors. Retries with exponential backoff should be reattempted until the retry counter reaches its maximum limit. However, super-fast retries end up causing retry storms and cascading failures.

In a microservices architecture, this pattern is very common; each service is hosted as a separate service and transient and communication errors are common. All the calls are made to the service over a network, and if there is a loss in network connectivity, it creates service timeouts and operation failures. These transient faults are usually self-correcting, and if the process is repeated after a while, it might be successful.

If the service is not responding and the consumer application receives an error, it can handle it using the following strategies:

- **Retry**: If the error is transient, make the call immediately to the service to get a success response. However, a bad implementation of such a retry can bring the whole service and many others down. For example, if you configure for retrying on, say, failure A 5 times, and the upstream service is having an issue, it will suddenly have 5 times the load and will have an even harder time recovering.

- **Retry with delay**: This will have some delay before making the next call to the failing service. The retry will happen based on the retry limit you have set.

- **Cancel**: If the error is not transient, it may cancel the operation since the operation might not be successful, even it is repeated, and there is no reason to keep the consumer waiting. In this case, it is good to fail fast and let the consumer know why the service has failed.

The following diagram depicts a scenario where the retry pattern can be used:

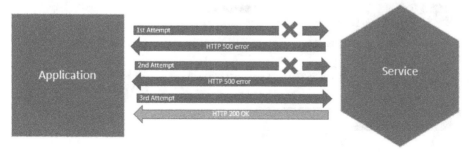

Figure 4.8 – Using the retry pattern between an application and a service

In the preceding diagram, an application is making a call to the service that is failing. The application implements a retry pattern to retry up to three times if the service fails.

The circuit breaker pattern

Sometimes, the errors are not transient and not related to any network connectivity or communication. This means that the fault is not temporary and requires effort to remediate. These faults could be related to service maintenance, redeployments, bugs in the application, or just a sudden spike in traffic because of end user behavior. Sending every request to undergo the retry approach is not a good practice. If the service is failing due to other reasons, it is elemental to fail fast and return a fallback response. This is where the circuit breaker pattern can be used.

Just like the retry pattern, the circuit breaker pattern can be implemented as part of the microservice's implementation. This pattern can also be implemented by frameworks, service meshes, or API gateways. It works in three states, as follows:

- **Closed**: It is like an electronic circuit – if the circuit is closed, the request will reach the target service. The initial state is a closed state, where all the requests are sent to the target service.

- **Open**: When a service fails to respond, the circuit becomes open for a limited time. In this state, none of the requests will reach the target service and fail fast. You can also return a fallback response to a consumer with a friendly message such as *"We are doing some maintenance, please try again after a while."*

- **Half-open**: The circuit becomes half open once the waiting time has elapsed in the open state. The first request will be allowed to be sent to a target service. If the service successfully responds, the circuit is closed; otherwise, it will open again for a limited time, as configured.

The following diagram depicts the circuit breaker pattern:

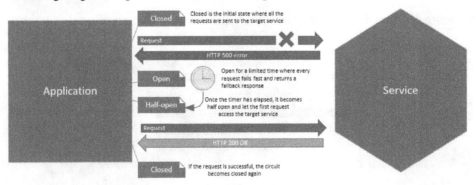

Figure 4.9 – Using the circuit breaker pattern between an application and a service

In the preceding diagram, when the first request was made, the circuit was closed. When the service responded with an HTTP 500 error, it opened the circuit for a specific amount of time. When the circuit was open, none of the requests were sent to a target service. Once the time had elapsed, the circuit became half open and allowed the first request to be sent out. Finally, when the service responded with an HTTP 200 status code, the circuit was closed once more.

You can also use the circuit breaker pattern in conjunction with the retry pattern so that when the circuit is closed and the request fails, it will not open the circuit until the retry counter reaches its maximum retry limit. The following diagram shows using the retry pattern in conjunction with the circuit breaker pattern:

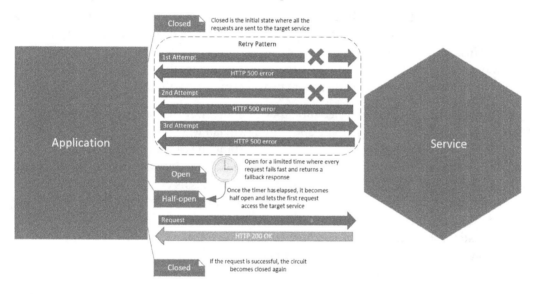

Figure 4.10 – Using the retry pattern in conjunction with the circuit breaker pattern

Various libraries allow you to easily implement the retry and circuit breaker patterns. The following table shows the most common ones that are used for .NET and Java applications:

Library	Platform	More Information
Polly	.NET applications	`https://github.com/App-vNext/Polly`
Netflix Hystrix	Java applications	`https://github.com/Netflix/Hystrix`

Let us next see how important scalability is when designing a microservice.

Scalability

One of the key advantages of microservices is that they enable businesses to scale out services separately. Companies transitioning from a monolithic architecture to a microservices architecture have an idea of how to update their systems, but they may struggle with what is usable or how it will benefit them if they implement it. A monolithic application running within a VM, for example, is not a better trade-off than microservices running within containers. The infrastructure team, who are well-versed in VMware or Hyper-V, may need to build skills in areas such as containers, Kubernetes, and others, whereas developers may need to build skills in terms of adopting different design patterns, container knowledge, and so on.

Cloud-native applications usually leverage the benefits of managed services in the cloud that offer good SLA and easy management. For example, if you plan to host your container inside an AKS instance, you can easily set up auto-scaling for your cluster or pods. Moreover, if you plan to host your app as an App Service that also provides auto-scaling, there are various other services that provide similar capabilities.

Apart from infrastructure, the service should be built in a way that supports scalability. There are many scenarios in which the application will fail when you scale out. For example, if you are building stateful services and keeping the cache within the container itself, they would fail when you scale out. Since the other container does not have access to that cache, the same state cannot be used, even if it creates its own local cache. To address this concern, you can, for example, use a distributed cache such as Azure Redis Cache or build services on Azure Service Fabric, which provide a complete framework of building stateful services. Another example is when you want to keep data such as logs or application-specific data private to the container. Scaling out the container will make the data distributive, resulting in inconsistencies and making maintenance harder.

Microservices are independent and self-contained. We'll look at why sharing dependencies is one of the drawbacks and how to build self-contained services in the next section.

Sharing dependencies using older techniques

In the n-tier architecture, each layer has a direct reference to the other layer, which makes it tightly coupled and dependent. With microservices, this is one of the anti-patterns. There are cases where one service needs to call another service to complete a transaction. In this case, the service is dependent on the other service. To overcome this challenge, you can build self-contained services.

Building self-contained services

Self-contained services help you collaborate with other services using patterns such as Saga and CQRS. Instead of making direct HTTP calls, the source service communicates with the other service over a message broker. The source service publishes a message, and the other service reads the message and responds. One approach is to keep two queues for requests and responses. That way, both services can communicate asynchronously over this channel.

The following diagram shows the collaboration between services over an asynchronous communication channel:

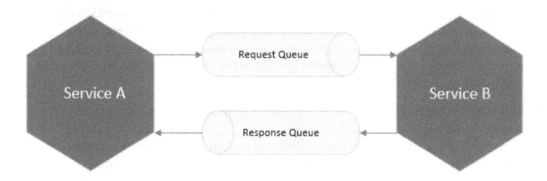

Figure 4.11 – Implementing request and response queues to establish asynchronous communication between services

If **Service A** wanted to get some data from **Service B**, it must publish the message to the request queue. Since **Service B** is subscribed to that queue, it will read the message, process it, and then write the response in the response queue. **Service A** needs to subscribe to the response queue so that when the response is published, it is received by **Service A**. Each service maintains two separate sessions to publish and consume messages from request and response queues. This approach helps reduce I/O resources and allows services to function in non-blocking mode, as well as serve a much larger set of requests in comparison to synchronous HTTP communication. Implementing this technique for all cross-service communication may not be the best option and should be considered based on the business scenario. For example, due to asynchronous communication, it may participate in multiple communications without assuring you that the sequence of responses is preserved in the order of the incoming requests. To overcome this challenge, you can add a unique identifier, also known as a correlation token, to each message, which can be used to correlate with the original message.

Another technique is to implement **Command Query Responsibility Segregation (CQRS)**, where **Service A** can keep a local copy of the data from **Service B** by catching it. This synchronization between two services could be done using an event-driven approach. If any change occur in the **Service B** database, it will publish a message to synchronize the **Service A** database. In this case, **Service A** can read the local database copy instead of making a call to **Service B**. To learn more about Saga and the CQRS pattern, please refer to *Chapter 5, Data Design Pitfalls*.

Reinventing is not an anti-pattern in microservices

Reusing and sharing the dependencies in monolith applications is a good practice. Many technological companies who develop software maintain their own frameworks to reduce their development efforts and reuse any functionality when needed. However, when transforming an application from a monolith into a microservice, reusing the existing component to build a new service is not only difficult but also restricts you from adopting new technologies and frameworks.

In microservices, reinventing the wheel is healthy. Companies cannot benefit from the technological advances of new technologies if they don't reinvent. Using the existing components of the framework with a monolith will restrain them from adopting newer technologies, which will reduce their ability to be agile and technology-agnostic.

Shifting from a monolithic application to a microservice model often allows technical teams to innovate at their own speed and scale according to their needs. Teams can experiment with innovative technologies to develop new services. The risk factor is reduced compared to a monolithic system because if one part fails, it does not impact the whole system.

Understanding such pitfalls is vital when developing or migrating to a microservices architecture. Taking all of these considerations into account will help you seamlessly transition from a monolithic application to a microservices architecture.

Summary

In this chapter, we learned about the importance of replatforming monolith applications into microservice architectures and how to decompose them. Transitioning from a monolithic architecture into a microservices architecture is not simple, and there are many factors and pitfalls to consider. Availability, reliability, and scalability are important factors when decomposing an application, and failing to give them importance is one such pitfall. Hence, we went through some concepts around availability and scalability, and we learned about the patterns that are used to build resilient microservices. Last but not least, we addressed the disadvantages of using the same components of monolithic applications and not reinventing the wheel to take advantage of emerging technologies that provide better alternatives.

With this knowledge, you can begin your journey of decomposing the current legacy applications into microservice architectures and reap the benefits of all the factors we've addressed in this chapter. The next chapter will focus on the pitfalls to avoid when building a microservice data architecture. You'll learn about common anti-patterns to avoid when constructing a data architecture and how different approaches can be used to ensure data continuity and help construct autonomous, flexible, and robust microservices.

Questions

1. Why should you migrate from a monolithic application to a microservices architecture?

2. What are the factors to consider when replatforming a monolithic application to a microservices architecture?

3. Why are availability, scalability, and reliability important and how can you implement them?

Further reading

For more information regarding what was covered in this chapter, take a look at the following resources:

*   *Practical Microservices*, by Umesh Ram Sharma

Section 2: Overview of Data Design Pitfalls, Communication, and Cross-Cutting Concerns

In this section, you will learn about data design pitfalls, communication pitfalls, and cross-cutting concerns, along with examples and alternative patterns to address these challenges.

This section comprises the following chapters:

5
Data Design Pitfalls

Data is the lifeblood of any business application. As developers and architects, we are constantly focused on our application's data and its state, how we transport that data, and how we persist our data. In traditional approaches, such as a waterfall, the first thing we design is the database. However, in my opinion, this is the incorrect approach to take. This is because, in the beginning, our knowledge of the problem space or business domain is naïve. In turn, this could lead to many columns in our database tables that are never used and schemas that are incomplete or bloated. Microservices best practice states that each service is responsible for its own data. We should begin with the development of our domain model and then determine which persistence strategy to employ. We should design our services and identify the aggregates that make up and form the boundaries around that microservice and its transactions. In this chapter, we will explore some of these strategies and understand where and when to employ them.

Indeed, we will explore some important pitfalls that can hinder the data development of your microservices. In this chapter, we will focus on the following topics:

- Keeping a single shared database
- Normalizing read models
- Understanding the **Command Query Responsibility Segregation (CQRS)** principle
- Not knowing how to handle transactions
- Not choosing consistency and concurrency
- Not knowing how to perform reporting

The pitfalls of keeping a single shared database

The number one mistake that teams make when deciding to go with microservices is trying to share a single database and scheme with multiple microservices. So, why is this such a bad idea even though it certainly seems easier? Well, let's start with the definition of microservices.

The definition of a microservice states that each microservice should have its own database or data store for which it is responsible for maintaining the state of its data. Therefore, it makes sense that we adhere to this principle of a single database when creating our microservices. It is also reasonable to conclude that when we break up a monolithic application into microservices, we must break apart the large monolithic legacy database into smaller, separate data stores of one per microservice. This breaking up of mature legacy databases is no easy task and should not be taken lightly. An activity such as breaking up a mature database into smaller databases requires much planning and preparation. We need to ensure that we have a deep understanding of the monolithic database and its relational data. Additionally, we need to procure the knowledge of the data that supports the service we are creating and separate that model from the legacy database into our new microservices database.

It is paramount that we break up the large monolithic database into smaller data stores for each microservice in order to loosely couple the separate services away from each other. If we fail to do this, our microservices will become tightly coupled. This means you will have simply created a large, distributed monolith that is hard to maintain and manage. Additionally, it means that we will have created more complexity for ourselves by separating the microservices and two separate code bases that are still being tied to a monolithic database. We've just made our world far more difficult to manage when it comes to microservices and have defeated the whole purpose behind them. With our microservices tightly coupled by the monolithic database, we will start to experience deadlocks and high contention situations in our database where multiple microservices serving many requests try to update a single record in the database.

You can take two different approaches in which to split the database and its data. One is to start with the database first, and the other is to start with the code first. The latter helps us to break the database up into smaller databases that align with the domain or business feature we are trying to create the service for. The code-first approach, I believe, is the safest, as we move the model from the old legacy database to the new service by building the code model first. By doing so, we can build our service in isolation with supporting unit tests to first prove our design.

We can achieve this by initially breaking out the code from the legacy application and designing the models that support the business features. Then, we can use a proxy to direct traffic to the new service as we test and learn about the data that is required for the service. Finally, we can compare and contrast this data and its model against the legacy database. By taking a code-first approach, we can decide how to proceed when separating the schema and its columns and properties from the legacy data store. Additionally, we can have greater confidence in using the unit test to ensure our assertions are correct. Earlier, we discussed the importance of a microservice having its own separate data store and being responsible for keeping that data in the correct state. But what if that data's needs would be better served if it was from a different persistent store that meets the needs of the data itself?

What if we need to improve retrieval performance whether it be in big data or within certain CQRS patterns, which we will discuss shortly? Some implementations of CQRS have a database of read operations for queries and another separate database of write operations in which to store data. The stored data is then replicated into the read database, called a materialized view for the read operations. The materialized view model has been optimized for reads and queries.

Relational databases were designed to optimize the storage of data in rows, mainly for transactional applications. A columnar database is optimized for the faster retrieval of columns of data and is widely used for analytical applications. Column-oriented storage for database tables is essential in analytic query performance because it considerably reduces the overall disk I/O requirements and reduces the amount of data we need to load from the disk.

What if we need a columnar database for high-performance aggregation or a document database for a materialized read view of our data? This helps in faster reads or a relational database to better facilitate durable transactions.

When we find our data sources and repositories coming from many different places along with different paradigms such as big data, relational databases, or document databases, then we need to explore polyglot persistence and what it brings to the table. We will do this next.

Embracing polyglot persistence

So, should we choose a NoSQL document database or a SQL relational database? Let's answer this question and more in this section. We will explore different persistence models and compare and contrast them to see how they pertain to polyglot persistence.

There has been much discussion around data persistence and whether SQL relational databases or NoSQL document databases are the way to go when it comes to persisting data and its state. As we continue our journey in the realm of data design, we are introduced to the emerging concept of polyglot persistence. Here, we can use several paradigms, from NoSQL to SQL relational stores, when dealing with data persistence and storage for our microservices.

Polyglot persistence is the approach of storing data in the best data store available to meet the needs of the application or use case. For instance, this could be a data lake of big data for the storage and retrieval of large raw data, a document database such as MongoDB (CosmosDB) for the faster reading and writing of JSON documents, or a tried-and-true relational database such as SQL Server. In some cases, you might even use all three and will require a custom design as there is no one tool that you can use to handle persistence across NoSQL or relational databases as far as I am aware.

In recent history, document databases have become particularly popular in the application development world. The reason for this is the common practice of passing data back and forth through RESTful web APIs and persisting data in the raw format of JSON documents without the need to map to database columns. This has led to numerous discussions and debates on the importance of choosing one or the other when it comes to relational databases or document databases. Both document databases (NoSQL) and SQL databases (relational) have their strengths and weaknesses. Indeed, in some cases, you will find that you need both a NoSQL store and a SQL store, and this can be achieved via a repository pattern. We can now provide different implementations of our data stores, either relational-based or document-based solutions, by abstracting away the details of how the data is persisted through interfaces, such as in a repository pattern. Polyglot persistence uses the concept of the best database model for the job at hand.

It is important to understand that not all data needs are equal. Some require a different approach and careful consideration. Polyglot challenges question what our data needs are, whether the old model works for our use case, and whether we should consider a different approach.

Modeling data to meet whatever paradigm we find ourselves in is paramount to good design, so we will explore different approaches such as a relational data store and a NoSQL document data store. Note that they are very different when it comes to modeling our data and each requires a different mindset.

Normalizing read models

In SQL relational databases, normalization makes perfect sense and should be implemented by creating a table for each entity. Additionally, we can have separate tables for all of the children's objects and use foreign key identity to map the relationship between those entities.

In a document database, this can be a detriment to performance and bad design, but there are some cases when we need to normalize document database models. The best practice is to embed everything within the parent entity; however, in the case where the embedded objects can grow without bounds, this can create an issue with document size. Therefore, in this scenario, we need to model our data with a reference just as we do in a relational world using identities.

Another example of the correct time to normalize a document model is when dealing with subordinate objects that change frequently. For example, we have a parent document called the customer, the customer has a list of stocks, and the stock prices are updated every 5 minutes. In our document model, we want to move stocks out of the parent, in this case, the customer, and into a separate document object called stocks, which is a collection of customer stocks. This is so that we don't have to update the parent (that is, the customer) every time we need to update an embedded object (that is, a stock) that has changed, such as a stock price. So, we will want to move the stock object into a separate document object and reference it with its ID. Then, we can update the stock itself and not the parent object that owns the stock object. For instance, we can view an example of the documents for a customer in which we track their stocks and update the prices of those stocks, as demonstrated in *Figure 5.1*. However, note that the golden rule in the document database world is when in doubt, embed, embed, embed.

The following code block provides an example of a customer document that uses an ID to reference the customer's stock document. The customer document also includes the name of the stock; this is because some queries might need to retrieve a list of customer stocks without having to send an additional query to the stock document to retrieve the names of the customer's stocks:

```
##### Customer Document With Stock Id and Stock Name
{
    "customerId": "cust123",
    "firstName": "Bob",
    "lastName": "Nelson",
    "Stocks": [{
            "stockId": "stk123",
            "stockName": "Socks Stock"
```

```
        },

        {
            "stockId": "stk234",
            "stockName": "Another Stock"
        }
    ],
    "isActive": true
}#### Stock Document
{
    "customerId":"cust123",
    "stocks":[
        {
            "stockName":"socksStock",
            "stockId":"stk123",
            "stockPrice":85.56,
            "isTraded":true
        }
    ]
```

The preceding code is a good example of a customer and stock document that shows the normalization of the customer document by referencing the stocks by ID.

Denormalizing read operations or keeping flat tables

A read model materialized view is a concept we find in document databases that facilitate high-performance reads. When designing models for reading and writing while dealing with read-heavy workloads, we want to optimize our JSON documents to provide the best performance for our queries. This is done by embedding all of the objects that are a part of that object's composition inside the JSON document; for example, with a person object that has an address object. In a relational database, we would have a table for the person and a table for the address. Then, you would join the person ID to get the address for the person and our query. In a document database, we would embed the address object inside the person object, as demonstrated in the following code. Then, when we query the person, we also automatically get their address without any join being required.

The following is an example of a person document with an embedded address object. In our example, this is what the JSON document will look like:

```json
{

    "firstName": "Bob",
    "lastName": "Nelson",
    "address:": {
        "street": "1st Street",
        "city": "Tampa",
        "state": "Florida",
        "postalCode": "33610"
    },

    "isActive": true
}
```

Now, let's move on to the next section, where we'll have an in-depth discussion on CQRS.

Understanding the CQRS principle

In this section, we will learn about CQRS, the different types of CQRS, and why we might want to consider this pattern. CQRS was introduced by Greg Young in 2010. It is based on the **command-query separation (CQS)** principle that was introduced by Bertrand Meyer, in 1988. In this scenario, we separate the responsibilities of querying or retrieving data from commands, which results in a change in state or mutates our data. The CQRS principle takes the concept of CQS and builds on it, adding more details and capabilities such as having a separate database for reads and writes. This can add complexity but has provided some useful capabilities such as persisting events known as event sourcing. We can play these events back to give us a very detailed audit trail or history of the state of our data: when it changed, who changed it, and why.

Types of CQRS

As mentioned earlier, CQRS was built on the principles of CQS. Additionally, we mentioned CQRS has added some additional capabilities and details. So, let's take a closer look at what those additional capabilities and details are. We will examine the different types of CQRS and the capabilities of each type.

A single database

A single database is the simplest type of CQRS. Here, commands use the domain and have a persistence layer that can employ an **Object-Relational Mapping (ORM)** for relational database stores. Another option is to also use a document database if you wish. ORM is not required, as we just persist the object directly into the JSON document, so no mapping is required. Queries can use a fast micro ORM such as Dapper for reads and queries and Entity Framework for persistence in the command stack.

Figure 5.1 illustrates a single-database CQRS implementation:

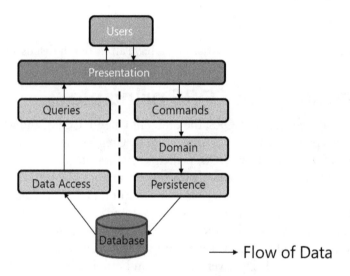

Figure 5.1 – A single-database CQRS

In the preceding diagram, there is a single database for both reads and writes. The data flows through the command stack and down to a single database. Then, it flows through the database, moving on to the **Data Access** layer. Finally, the data flows out through the **Queries** service layer and back to the user.

Two databases

The two-database approach uses two databases – one for reads and another for writes. This approach employs eventual consistency and introduces some additional complexity because we now need to keep both the read and write stores in the correct states. With this approach, the read performance is increased as we can also use a document database for a materialized view that is optimized for reads. We can keep our command stack clean and efficient and help concurrency with a relational database for writes to enforce transactions. It is common for an application to be more read-heavy, so by separating the reads and optimizing our read stack for queries, we can increase performance in the **Presentation** layer. *Figure 5.2* illustrates the two-database approach and how it uses a stored procedure to update a read after every update. Another approach to keep the read data store in the correct state is by inserting a service `insert` or `update` to update the read store. While this also introduces complexity, in some cases, it is a good option:

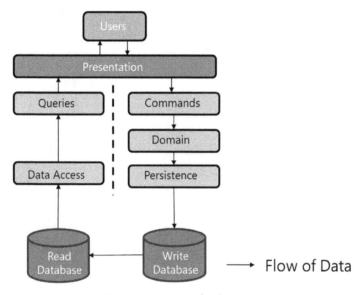

Figure 5.2 – A two-database CQRS

The preceding diagram illustrates the two-database CQRS implementation. We can see that we have a read database and a write database. The data flows through the command stack and down to the write database. Then, it flows through the read database, moving on to the **Data Access** layer. Finally, the data flows out through the **Queries** service layer and back to the user.

Event sourcing

Event sourcing is the third CQRS approach. It is the most complex of all the CQRS types in which we store events as entities using an event ID. We can replay events to get to the current state and to view the audit trail or history of the event entity. We modify the event entity when there is a state change and events become the source of truth. Just as in the two-database approach, we need to update the read database; this will be a materialized view of the last event and the current state of that event to optimize reads. Events can be replayed, but this operation is expensive. Therefore, we need a materialized view for reads. Event sourcing is very complex and should only be considered when the need for auditing the history of events, such as in healthcare or government compliance, dictates this need. We can conduct a point-in-time reconstruction of the state and view why it changed, how it changed, and for debugging purposes. This ability to be able to replay events can help meet the audit and compliance requirements of regulated industries and should only be considered when the use case requires such a feature due to the complexity and cost involved in event sourcing. Simply put, if you don't need it, then don't use event sourcing.

The following diagram illustrates the event sourcing CQRS implementation:

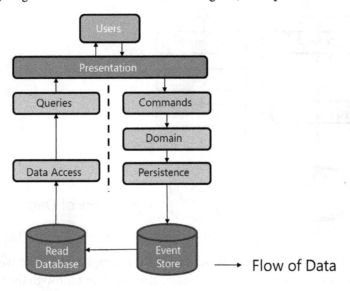

Figure 5.3 – Event store and read database

Here, we can see that we have an event store and a read database where data flows through the command stack. The data flows down into the event store where the event is persisted. Then, the current state of the event is mapped into materialized view, which has been optimized for reads. This materialized view is then added to the read database. Note that this can be done on the database layer or via a service layer that writes the materialized view to the read database. Finally, the data flows through the **Data Access** layer, out through the **Queries** service layer, and back to the user.

CQRS and microservices

A more common approach uses a single database with separate schemes and data models – one for reads and one for writes. This single-database approach of CQRS is more commonly found in microservices. The separate model for queries and the separate model for commands are known as view models and domain models. **ViewModel** is for queries that are used by client applications to view the internal aggregates, and the domain model is used by commands to mutate or change the data's state. This separation of responsibility between the two models keeps the queries independent from business constraints and invariants, which are derived from domain-driven design patterns that apply to the transactional boundaries of the root aggregates and the updates made to the root's internal objects.

ViewModel is used by clients for queries as predefined **data transfer object** (DTO) classes, which, if you will, serve as a contract regarding how data is presented to calling application clients.

DomainModel enforces business invariants and business rules that are part of the aggregate's transactional boundary.

The problem we are trying to solve with CQRS

The problem we are trying to solve with CQRS is that, in traditional architecture, the same data model is used to query and update the database. This simple approach works well for basic CRUD operations. However, as the application grows in complexity, this approach becomes difficult to manage.

For example, on the read side of things, the application might perform many different queries and return DTOs with different compositions and sizes. This can make object mapping complicated and tedious. The `write` side of the model can implement complex business rules and validations. As you can see, this can result in an overly complex model that is trying to do too much. This is where we need to clean things up and separate the concerns of mutation and querying.

Often, there is a mismatch between the `read` and `write` representations, and we can end up with properties or columns that we never use or implement. We might even need to update the properties despite never using them in our operations. This issue with data contention can raise its ugly head as two operations try to access or update the same data in parallel.

Figure 5.4 illustrates the traditional stack in a layered application:

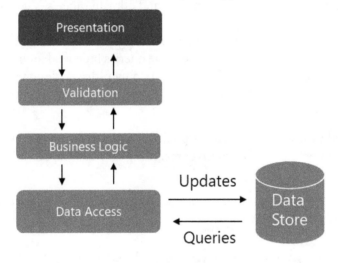

Figure 5.4 – The problem with one model for queries and updates

At a first glance, this appears to be a fine example of a separation of concern; it is and works well in many cases. However, while this might work in simple applications, as the complexity of the application grows, mixing methods that mutate the state along with the responsibility of returning a ViewModel adds even more complexity and breaks the single-responsibility principle. As you can see, as the request comes from a client, the flow of data traverses each layer. Therefore, each layer is responsible for reading and writing not only the mutation of the state but also requiring logic and mapping from a ViewModel back to a UI within the response. This can result in slower page loads, as the application must complete the update and return the response in the same request or transaction. To make matters worse, business logic can start to spill over into other layers. Additionally, business logic shouldn't be concerned with data access concerns or even be aware of it. Business logic should only be concerned with business rules, policies, and invariants rather than with the logistical concerns of data access.

Figure 5.5 depicts a bidirectional request and response schema where we request a change to the state, and at the same time, we need to return the view back to the UI, which makes us tightly coupled with the UI implementation. This can hinder performance under load due to the locks on the objects being updated and requested at the same time and then having to return a response:

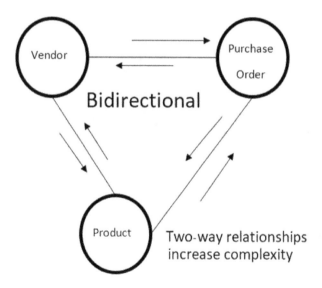

Figure 5.5 – Bidirectionally adds complexity and is hard to manage

The solution

CQRS helps us solve the problem that we face with one model by separating reads and writes into two different models using commands to update the data and queries to read data. This separation gives us a far cleaner way of retrieving data so that we can optimize our queries for better performance and keep business logic out of our queries.

The command should be tasked based on business logic and invariants rather than being data-centric. Additionally, the command should flow one-way, which is also known as unidirectional, as demonstrated in *Figure 5.6*. The command never returns anything to the caller, which facilitates this one-way flow. Commands can be placed in a queue and processed asynchronously rather than synchronously. Queries never modify the database; they return a DTO and do not concern themselves with domain knowledge. The models can be isolated, as depicted in *Figure 5.6*, where you can view a model for writes and a model for reads in the database and the direction of the flow for each responsibility:

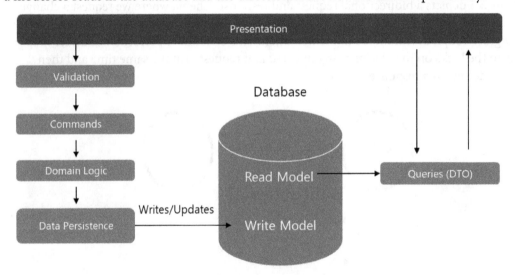

Figure 5.6 – Commands and queries have separate models

As we have gathered, in *Figure 5.7*, the one-way unidirectional approach is where we have a one-way request for the mutation of the data:

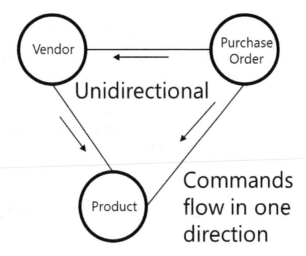

Figure 5.7 – Commands flow in one direction and mutate data

This makes commands much cleaner, and we can focus on business logic. Besides, we can place commands in some messaging frameworks such as Azure Service Bus, Kafka, or RabbitMQ and handle those commands asynchronously.

Challenges when implementing the CQRS pattern in our microservices

One of the challenges we face when implementing a pattern such as CQRS is **complexity**; for instance, when we implement event sourcing. Although the basic concept of CQRS is simple, it can lead to a more complex design; for example, the two-database approach and, as mentioned before, event sourcing, which has a very steep learning curve.

Messaging, although not necessarily a part of CQRS, is common to use. Messaging frameworks support commands and publish and update events such as with domain events. Messaging can add additional complexity, as we now need to deal with message failure and duplication. Alternatively, in some cases, we also need to ensure messages are processed in a particular order.

Eventual consistency refers to when we separate the read and write database. In this scenario, we now have the additional complexity of keeping our models in the correct state and ensuring that our read data does not become stale. We have to ensure that any changes to the write model are reflected in the read model. We have now explored CQRS and demonstrated how this pattern can either clean up or design data to help separate the responsibilities of reading and writing data. Additionally, we have examined how we can leave commands to mutate or update our data. Now, we need to understand transactions and the importance of maintaining the critical state of our data, and the role of transactions within this process.

The pitfalls of not knowing how to handle transactions

One pitfall that teams face is not knowing how to handle transactions that are distributed across multiple microservices. We will explore how we can frame transactions within a microservice and how to handle them when they involve multiple microservices. We will start with a concept that we explored in *Chapter 3, Microservices Architecture Pitfalls*, where we discussed domain-driven design and the concept of aggregation.

Aggregates are transactional boundaries. They form boundaries around transactions that enforce business rules and business invariants. These are then grouped by relevant entities that are part of the aggregate and its transactional boundary. These transactions are a single unit of work and can be made up of multiple operations that live in the space of that isolated transaction and produce a domain event that represents a state change to the aggregate and its internal entities. A command is an operation that performs the state change and causes the event to occur or be triggered. These events can notify the next step, or the system at large, of the success or failure of that transaction. If successful, they can trigger another command to be executed, and that command can trigger an event. This process will continue until all transactions are completed or rolled back, depending on the successful or failed completion of the transaction.

Transactions have defining characteristics, and they must be **atomic, consistent, isolated, and durable (ACID)**. Let's examine these characteristics as they pertain to microservices:

- **Atomicity**: This is a unified set of operations that must be successful or unsuccessful.
- **Consistency**: This means that the state is changed from one valid state to another valid state.

- **Isolation**: This helps to guarantee that transactions are concurrently executed in order and that they produce the same state changes those previous transactions produced.

- **Durability**: This ensures that the state changes are committed and executed regardless of a system failure or transient error.

While having a single database for each microservice is desirable, it also presents some challenges with the transaction when we need to keep the entire system in a valid state. For instance, in a monolithic world, this is a much easier task. Because the data is distributed, this presents the challenge of how to ensure the consistency of the entire system. This can be achieved by providing a mechanism for orchestration and managing these transactions across boundaries. Saga is an appropriate pattern to help orchestrate and manage this transaction and conduct any compensating transactions if a rollback is required.

Not choosing the right consistency and concurrency

As we have gathered, it is important to choose the right consistency and concurrency approach when handling distributed transactions.

A distributed transaction saga pattern can be used to manage distributed transactions across services or boundaries and maintain consistency across these boundaries. A saga is made up of an ordered sequence of transactions, whether they are updates, deletes, or inserts. If one of these transactions fails, the saga is responsible for rolling back any of the changes made. This process is referred to as a compensating transaction, which nullifies any previous transactions to return the system to the correct state.

Figure 5.9 shows a high-level view of a saga and its encapsulated responsibilities:

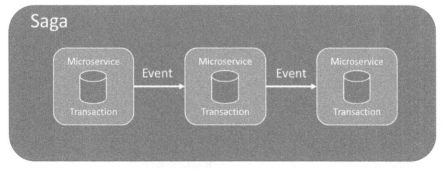

Figure 5.8 – A high-level saga

Each transaction happens in isolation locally in each microservice. Each microservice changes the state of its respective data via a command and then raises an event that can place a message on a queue to be picked up by the next service. Then, that service executes its isolated transaction, and so on, down the line. If the transaction fails, then the saga must execute a compensating transaction. This returns the preceding data store to its previous correct state. Both commit and rollback strategies can be achieved by using one of two implementations, that is choreography or orchestration.

The choreography implementation of a saga

Choreography does not employ a centralized control mechanism. Instead, each service either places a successful message and the next service picks up the message and continues the distributed transaction, or it places a message of failure in which the other concerned services pick up and then roll back the data state based on the failure.

Figure 5.9 demonstrates the choreography implementation using a message broker, where messages are placed for services to process their transactions against:

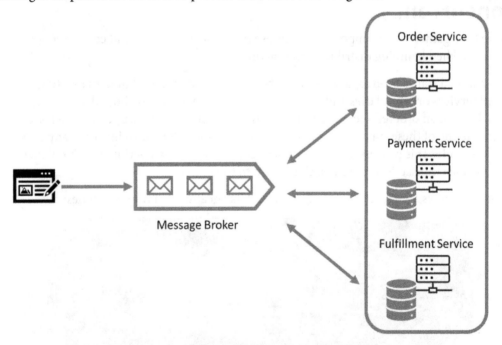

Figure 5.9 – Choreography using a message broker

We will now explore the benefits and challenges of choreography.

The benefits of choreography are as follows:

- It is excellent for simple workflows that only require a few services and do not need complex coordination rules or logic.

- It doesn't require supporting services for implementation or maintenance.

- It doesn't have a single point of failure as the services are required to perform rollbacks and complete the transactions based on the messages they receive.

The challenges with choreography are as follows:

- When additional steps are required, the workflow can become muddled and confusing.

- Cyclic dependency can crop up between services due to the fact that they consume each other's commands.

- The testing of integration can become difficult due to the need for all services to be running in order to properly test.

The orchestration implementation of a saga

Orchestration employs a centralized control model that tells the services what transaction to execute based on a specific event. Orchestrators are durable as they store each request step's input and output, and then determine the state of each task and whether it was successful or not. Then, it handles any failures, retries, and recovery operations. The orchestrator manager is responsible for compensating transactions and ensuring a successful rollback is conducted if required. Complex coordination logic is maintained in one place and services are decoupled, as the only communication is through the process manager. The process manager is the orchestrator and acts like a state machine— maintaining and updating state on services in the workflow throughout the lifetime of the transaction.

Figure 5.10 illustrates the orchestration implementation of the saga pattern. As we can gather, a request comes in and the orchestrator manages the transaction:

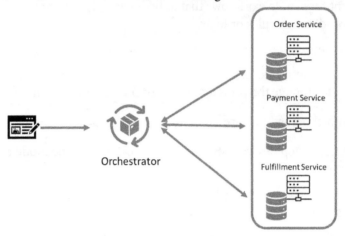

Figure 5.10 – The orchestration implementation of a saga using an orchestrator to manage transactions

Now, let's take a look at the benefits and challenges of the orchestration implementation of a saga pattern.

The benefits of orchestration are as follows:

- It is excellent for complex workflows that require many services to be coordinated along with complex coordination rules or logic.
- It is best for a situation when a high level of control is required for overall service transactions.
- Services do not need to be aware of the other services commands and their transactions, which greatly simplifies any business logic.

The challenges with orchestration are as follows:

- An added complexity is introduced due to the necessary coordination logic.
- An additional point of failure is due to the orchestrator's management of the entire workflow.

A saga pattern helps to ensure data consistency in distribution and reduces the coupling of the service's rollbacks or compensations for any failed transaction and helps keep our services in the correct state across the system.

Now that we have demonstrated how to maintain the state of our services across systems, we need to discuss how we can monitor the state of our data, produce reports, and conduct analysis of the data in our system as a whole.

Knowing how to perform reporting

Reporting is important to the business; to be able to analyze and monitor business outcomes, we must have a good reporting strategy. In most cases, reporting with monolithic applications and databases is a straightforward process. All we do is write stored procedures with complex joins and aggregates on the data to create views of the data. This is not possible in a distributed world; we do not have access to joins as the data is physically separated. So, how do we carry out these complex joins in a distributed microservice architecture? Well, the answer to that question is that we carry out the joins on the application layer in memory, not on the data layer. These joins are conducted in a reporting service using an API composition pattern.

The API composition pattern

A composition pattern calls each of the respective query services of each microservice that queries the database. Then, the API Composer service executes the joins on the data in memory from the disturbed services, combining all of the data inside the appropriate view for reporting purposes. CQRS makes this easier, as we have already created a query service that is responsible for querying a materialized view of the data in that service's read model or read database. Then, the API Composition services call our query stack within the CQRS service and combine all of the data from the distributed services, as demonstrated in *Figure 5.11*. The major disadvantage of this approach is that when you are dealing with large datasets, performance can take a hit. Some of this can be alleviated by creating a reporting database to persist the results of the queries in the reporting database and then conduct joins there:

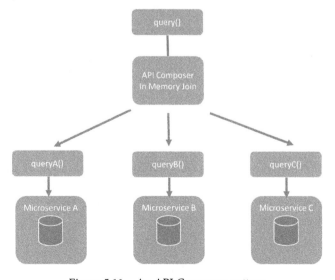

Figure 5.11 – An API Composer pattern

The preceding diagram illustrates the API Composer implementation, where an API Composer sends queries to the owning microservice of the data that is needed. Then, it combines the data received from those queries together in memory to provide that data to a ViewModel or for a report.

Summary

We have explored some important pitfalls that can hinder your microservice's data development. These include the pitfall of keeping a single shared database, where we learned that we should separate the data from a monolithic database and place it in each microservice on a separate database. The responsibility for the copying of the data will be on the owning microservice. We discussed some strategies to break the data out of a monolithic database via a code-first approach. This is done by developing a model in code first and writing a unit test to prove our assertions. Then, we move the data out of the legacy monolithic database. We also discussed the database-first strategy, which can be more complex and present a higher risk.

Additionally, we examined the pitfall of normalizing read models. Here, we learned that we don't always have to normalize the read model as we do in document databases. Document databases provide high-performance querying partly due to the embedding of children objects inside the parent object. This allows us to back the data we need with a key-value lookup without the need for complex joins such as in a relational database. These objects are made up of JSON documents that are stored directly inside the document database.

We examined the pitfall of not knowing how to handle transactions. We learned that we need to understand the boundaries around our transactions using the concept of aggregates. This is achieved by identifying the entities that make up a transaction that supports business rules and invariants. We learned about the autonomy of a transaction. We examined how to execute atomic transactions across a distributed architecture using our next pitfall, which discusses sagas. We explored the pitfall of not choosing the correct consistency and concurrency in the context of microservice architecture. Its distributed nature can create havoc with our data concurrency and consistency. The data involved in a transaction can span across the entire application that the microservices make up. We discussed how we can use a managing mechanism such as a saga to execute those distributed transactions and ensure all services involved in the transaction are completed successfully and that the data of each service is in the correct state. If the transaction fails, then the saga will execute compensating operations to roll the data back to the previous state.

Finally, we explored the pitfall of not knowing how to perform reporting by implementing an API that is responsible for building complex reports for business analysis. This API is called a Composer API, and it combines the data from separate services into a report model and joins the data together in memory.

Now, we can move on to consider our data design and examine some tools in which to address the challenges we will face on our journey to a microservice architecture.

In the next chapter, we will discuss communication pitfalls and prevention. In particular, we will explore communication between microservices across process boundaries.

Questions

1. Transactions must be atomic. Is this true or false?

2. Polyglot persistence uses one data store to service your data needs. Is this true or false?

3. What are the different types of CQRS?

6
Communication Pitfalls and Prevention

Microservice architectures are implementations of distributed systems, where microservices are designed to be fine-grained, loosely coupled, and autonomous in order to deliver a business capability. These microservices communicate with each other across process boundaries, which are hosted across the network, to address different business use cases. The network communication is unreliable; hence, microservices should incorporate the necessary logic to enable effective communication across service boundaries to build resilience. The challenges of designing a distributed system are well known. They are described by Peter Deutch as assumptions developers make while they transition from a single-process application to a distributed system. These assumptions are defined as fallacies of distributed computing. The fallacies are as follows:

- The network is reliable.
- Latency is zero.
- Bandwidth is infinite.
- The network is secure.

- The topology doesn't change.

- There is one administrator.

- The transport cost is zero.

- The network is homogeneous.

In this chapter, we will discuss the different aspects of communication that enable information sharing between different components of microservices architecture to deliver business value. We will also discuss the challenges of different communication approaches and how we can address them using solutions. In this chapter, we will be covering the following topics:

- The fundamentals of microservices communication

- Different architectures and their communication styles

- Direct client-to-microservice communication

- The overly complicated API gateway

- Request timeout

- Long chains of synchronous calls – the retry storm

- The dark side of event-driven microservices

- Avoiding service versioning

- Service meshes

- Dapr versus service meshes

By the end of this chapter, you will understand the varying ways in which microservices communicate with each other, their challenges, and how to address them using different approaches. In addition, you will learn about the importance of versioning and its role in the creation and evolution of microservices.

The fundamentals of microservices communication

Before we start our discussion of the different communication challenges, first, we will explore the fundamentals of communication in a microservices architecture. In the next few subsections, we will briefly discuss the core concepts of communication, as they form the basis of a microservices architecture.

Synchronous versus asynchronous communication

To enable communication between clients and services, we will explore the fundamental differences between various alternatives to help us choose the right communication type for our microservices. In **synchronous communication**, the client sends a request and waits for a response from the service. Synchronous communication can either be blocking or non-blocking, but the client can only continue further processing once a response has been received from the service. This pattern is also known as the synchronous **request/response pattern**. In service-to-service communication, request/response patterns can either be synchronous or asynchronous. In async API calls, the call is made, but the calling service is free to do other processing while the other service is working on a response. This pattern is useful in improving the throughput of the service.

In **asynchronous communication**, the client sends a message to the message broker, which acts as a middleware between client and service. Once the message has been published to the message broker, the client will continue processing without waiting for a response. The service can subscribe for specific messages to the message broker and place responses back in the message broker after processing. Both client and service are unaware of each other and can continue to evolve without any dependency.

In most cases, a microservices application uses both synchronous and asynchronous communication to address different use cases. However, if possible, asynchronous communication is the preferred approach, as it offers loose coupling and independence between microservices via messages. In the following section, let's explore messages and their types.

Messages and message types

A message is defined as data that is sent between the client and its receiver, which serves as a means of communication by exchanging information. There are different types of messages that are categorized based on their intent. These message types are described as follows:

- A **command** is defined as a message sent by the client with an expectation that the receiver performs an action.

- A **query** is defined as a message sent by the client with the intent of expecting a response from its receiver. The response could be a success or a failure. If no response is received, the client can decide to retry the request.

- An **event** is defined as a message that indicates an event that happened in the past in the system without any expectation from the receiver. The receivers of the event can ignore or perform necessary actions based on the information provided in the event.

Asynchronous message-based communication

In event-driven communication, services communicate with each other via events. Any state change of a microservice is presented in the form of an event that allows interested parties to subscribe to these events and perform necessary actions. This pattern is known as **publish/subscribe**, where a service can publish messages to a message broker and one or more consumers can subscribe to process those messages. Synchronous messaging can be converted into asynchronous messaging by introducing a queue that guarantees that each message is processed by one and only one consumer. This approach helps to decouple services, allowing them to change and scale independently.

Having understood the fundamentals of microservices communication, let's move on to discuss different application architectures and their preferred ways of communication.

Different architectures and their communication styles

In this section, we will understand the difference between intra-process and inter-process communication. In addition, we will learn about the similarities and differences between different architectural styles with respect to communication and how they use the two communication mechanisms for information sharing.

Intra-process communication is a mechanism that enables a process to share data between two threads using in-process memory, while inter-process communication is a mechanism that shares data between two isolated processes, either using messages or a **remote procedure call** (**RPC**). Both communication styles are well supported, though it's important to understand how different architectures drive the dominance of one communication style over other in an application. For example, the distributed system promotes the idea of inter-process communication over intra-process communication. However, any communication between components inside the service should be intra-process. Let's explore this, in more detail, with the help of the following diagram:

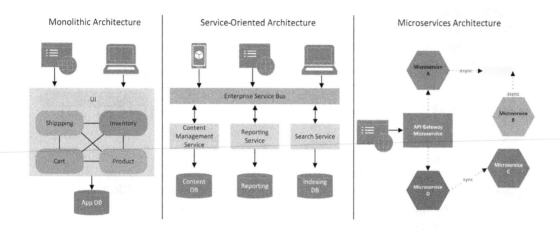

Figure 6.1 – The monolithic, SOA, and microservices architectures

In a **monolithic architecture**, the application is built as a single unit and hosted as a single process on a server. The monolithic application is composed of different components that are interconnected and interdependent in nature. Both synchronous communication and asynchronous communication are used in monolithic applications, but mostly, you'll find that the communication between application components is synchronous, while the communication with the user interface and external systems is preferred to be modeled as asynchronous.

The **service-oriented architecture** (**SOA**) is a company-wide approach that is used for building and integrating new and legacy systems using communication protocols. The communication between these services is performed using an **enterprise service bus** (**ESB**), which is responsible for message validation, routing, and transformation. In SOA, more emphasis is placed on integration and less emphasis is placed on the modeling of these services.

Microservices are an implementation of the distributed system, which promotes the idea of inter-service communication (that is, inter-process communication) between microservices. Simply put, inter-service communication can be categorized as synchronous and asynchronous communication. The synchronous communication can be achieved with lightweight generic protocols such as REST or gRPC, while asynchronous communication is supported via queues and message brokers. It's recommended that you use domain-specific protocols to handle inter-service communication, for example, the **Real-Time Streaming Protocol** (**RTSP**) for video streaming. In *Figure 6.1*, microservice A and microservice B communicate via messages using the queue, while microservice D and microservice C communicate via an RPC.

When opting for a distributed architecture, it's important that you understand the differences between the different protocols that can enable inter-process communication between various services. In the next section, we will discuss the various ways in which clients can communicate with microservices.

Direct client-to-microservice communication

In client-to-microservice communication, the client talks directly to the microservice. This approach is simple to implement, but it introduces tight coupling between the client and microservice, which makes the client more susceptible to breaking changes. Breaking changes are highly discouraged and should be avoided to ensure backward/forward compatibility. Further, the client needs to be aware of the business use cases and how it can address them by calling different microservices that result in violating the domain-driven design. Additionally, each microservice call will result in additional latency that affects user experience and mobility. Exposing multiple microservices directly to clients will expand the attack surface and will make the system and organization susceptible to attacks. Another challenge with direct communication is the synchronous calling of services, which undermines the benefits of microservices. The following diagram depicts a direct client-to-microservice communication pattern:

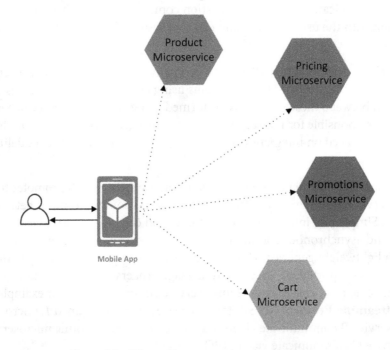

Figure 6.2 – Direct client-to-service communication

In the preceding diagram, the mobile application communicates directly with e-commerce microservices by directly invoking individual microservices. The user journey starts by browsing the product listing, selecting products, applying for promotions, and, finally, performing a checkout. As part of the communication, the client needs to perform multiple round-trips to each microservice, which results in additional latency and affects the user experience. To address these challenges, it's recommended that you consolidate user requests using an API gateway pattern. Let's explore that in more detail next.

The API gateway pattern

The API gateway pattern introduces a reverse proxy between the client and the microservices. It acts as a single point of entry for all client requests. API gateways are an implementation of API gateway patterns, provided by different vendors that are responsible for redirecting client requests to microservices for further processing. It also decouples client apps from internal microservices to enable service evolution and refactoring for internal microservices without causing any breaking changes to client apps. It's a good idea to keep the API gateway close to the microservice to reduce latency. The API gateway is a single point of failure, and it should be protected with high availability. You can use Azure API Management to expose APIs, protect your clients against interface changes, enforce security by adding various policies, and much more. You can learn more about the API gateway in *Chapter 7, Cross-Cutting Concerns*.

So, you should now have a basic understanding of how to address the challenges of direct client-to-microservices communication using an API gateway. Next, we will explore how choosing a single API gateway can result in creating ambiguities and the overall maintenance of the API gateway.

The overly complicated API gateway

A single API gateway might be ideal for a small microservices application, but it can turn out to be overly ambiguous if you are dealing with many microservice with various consumers. The following is a list of challenges that are often discussed while exposing APIs using API gateways:

- Each consumer has different information requirements from an API.
- Each consumer has different latency requirements.
- Protocols are supported by each client.

Also, API gateways are implemented as a mechanism to orchestrate backend microservices. The practice, well known as API gateway, is aggregator microservice. This concept is explained, in detail, in *Chapter 3, Microservices Architecture Pitfalls*. In the following diagram, a single API gateway is implemented to address the various requirements from different consumers:

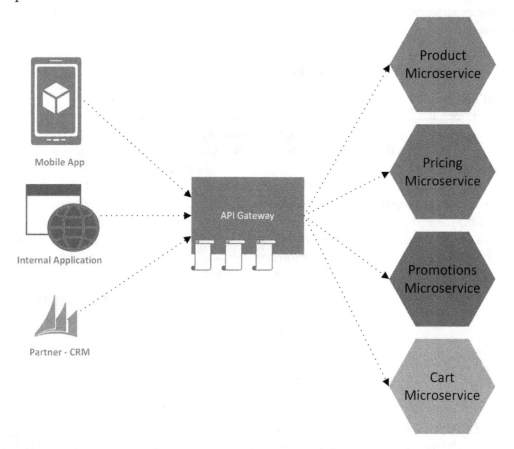

Figure 6.3 – A single monolithic API gateway

The API gateway also acts as an aggregator microservice, which brings the orchestration logic to the API gateway and, therefore, making it more complex and difficult to maintain. These API gateways can be a potential bottleneck and a source of single points of failure for the overall system. However, multiple API gateways can help to address the needs of different clients. Further, aggregator microservices can be used to orchestrate different microservices or materialized view patterns, which can help to reduce the communication between microservices. Let's explore them, in detail, next.

Multiple gateways – backends for the frontends

To keep an API gateway implementation simple, you can have multiple API gateways based on different types of consumers. Sometimes, it makes sense to have separate API gateways for altogether separate purposes; for example, one for external clients, one for internal service calls, and one for partners to consume APIs, where each API gateway contains a different set of policies and subscription model. It also enables developers to provide optimal APIs for each consumer by addressing their data and latency requirements, reducing round-trips, and improving user experience. Another benefit of this approach is that it segregates APIs based on their security and risk profiles. Furthermore, it helps you to address the challenge of a single API gateway becoming the bottleneck and a single point of failure for the system. Many cloud providers offer API gateways as a service to help cloud-native developers use these services to publish their APIs without needing to reinvent the wheel. Authentication, authorization, service discovery, rate limiting, throttling, load balancing, logging, tracing, correlation, IP whitelisting, and response caching are just some of the features supported by API gateways. The following diagram depicts multiple API gateways, each acting as a backend for the frontend:

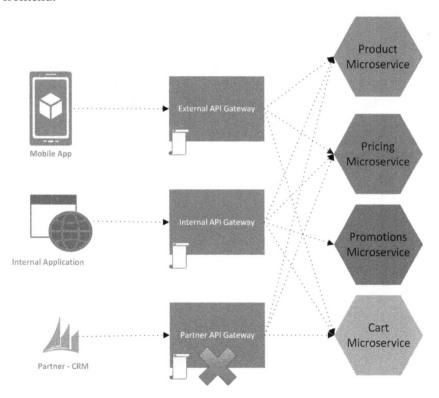

Figure 6.4 – Multiple API gateways

In the preceding diagram, three different API gateways have been implemented to increase the performance, availability, scalability, security, and maintainability of the overall application. Each API gateway has implemented consumer-specific policies to enable communication between microservices and their consumers through the API gateway. As shown in *Figure 6.4*, the partner API gateway is not available due to deployment failure, but this does not affect other API gateways, and other channels will continue to function without any disruption. This helps us to reduce the blast radius and reduce the number of affected consumers.

We can further simplify the communication between the API gateways and backend microservices by introducing an aggregator microservice. The aggregator microservice is responsible for orchestrating communication between backend microservices and API gateway. This concept is explained, in more detail, in *Chapter 3*, *Microservices Architecture Pitfalls*.

Materialized view patterns

The aggregator microservice has played an instrumental role in reducing the communication between the API gateway and backend microservice, but it has only helped us to shift the problem from the API gateway to the aggregator microservice. The materialized view pattern helps you keep a local copy (that is, a read-only model) of the data of other microservices, which is necessary for the primary microservice to function independently. This pattern eliminates calls to other microservices to increase the response time and reliability of microservices. The following diagram shows one attempt to make the overall microservices architecture more resilient by introducing multiple API gateways and materialized view patterns:

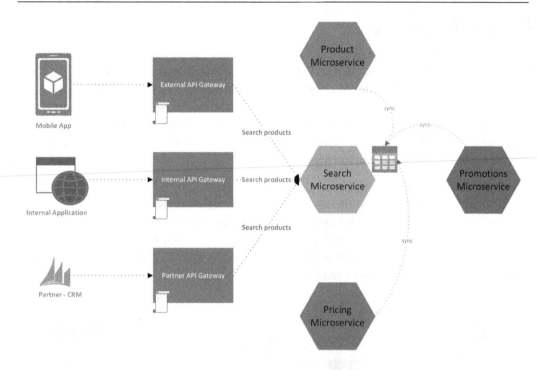

Figure 6.5 – Multiple API gateways with materialized view patterns

In the preceding diagram, we have multiple API gateways addressing high availability, scalability, and fault tolerance within the API gateway layer. The search microservice has a read-only model of the relevant data from multiple microservices, which is necessary to perform its operations. It's recommended that you asynchronously update the read-only model using queue or pub/sub patterns. For write operations, you can opt for either synchronous or asynchronous communication; though, asynchronous communication is preferred. The concept is explained, in more detail, in *Chapter 3*, *Microservices Architecture Pitfalls*.

Now that you have a good understanding of how to prevent your API gateway from getting overly complicated, let's move on to discuss the factors to bear in mind when enabling service-to-service communication.

Request timeout

In a microservices architecture, services are hosted as separate processes to communicate with each other over a network via RPCs. These microservices exchange messages using different protocols in order to consume interfaces exposed by these services. In synchronous communication, the consumer needs to establish a connection with a service to consume its capabilities. The ability to connect to a service to invoke its capabilities is defined as the **availability** of the service, while the time a service takes to respond to its consumer is known as its **responsiveness**.

Both availability and responsiveness are important aspects that need to be addressed by the consumer to enable resilient inter-process communication between services. The consumer of the service can choose to wait indefinitely or specify a timeout to wait for the response before it decides the next action. Although specifying a timeout seems like a good choice, it could result in a timeout anti-pattern. The following diagram depicts an inter-process communication between two microservices:

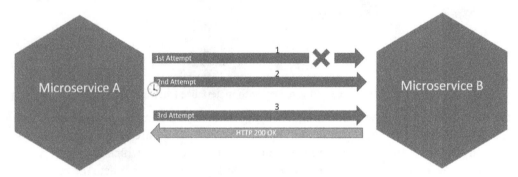

Figure 6.6 – Inter-process communication between two microservices

In the preceding diagram, Microservice A is trying to consume the capabilities that are available in Microservice B. Microservice A has implemented retry logic to re-communicate with the target service in the case of a failure. Additionally, it has implemented the timeout functionality to ensure service responsiveness. The timeout will inform Microservice A about the unavailability of Microservice B in the event of a failure. The series of events between the two microservices are explained as follows:

1. Microservice A tries to invoke a functionality that is available in Microservice B. However, Microservice A is not able to connect to Microservice B. In this case, Microservice B is considered unavailable for Microservice A. Microservice A can respond back to the client with an error message stating the unavailability or it can retry. There are multiple ways in which to handle this failure. To avoid synchronous calls, you can opt for a materialized view pattern or a publish/subscribe pattern, as we will discuss later in the chapter.

2. Microservice A will keep on trying to invoke Microservice B due to the *Retry Logic* implemented as part of Microservice A. If no response is received in the allotted time, this results in a timeout. In this case, Microservice B is considered unresponsive. If the timeout is not specified, Microservice A will continue to wait for Microservice B, resulting in a waste of resources.

3. In an ideal situation, Microservice A invokes the functionality that is available in Microservice B and receives a response in the allotted time.

At a first glance, setting up a timeout value seems to be the obvious choice, but finding the right timeout value that addresses the challenges of high volume, network latency, service unavailability, or overcommitted resources is difficult. For example, you could request to withdraw $5,000 of cash from an ATM. The last thing you want to happen is to get credited from the account and not receive any cash from the ATM due to a service timeout. The service timeout could happen due to network latency or a service load. You can try again, but you will find out that the account has already been credited. One approach to address this problem is to set a high value of timeout to allow these services to respond in times of high volume and load. A common way to address this problem is to set the timeout value to double the response time during peak hours, as specified in the following:

Average Response Time = 1500 ms

Response Time at Peak Time = 4500 ms

Timeout Value = 4500 ms x 2 = 9000 ms = 9 s

You might have wondered why this is an anti-pattern. Now every service must wait for 9 s to let its consumer know about the failure of a request. During this time, the client can either close the window or hit the submit button multiple times. This is not a good experience and should be addressed with a better strategy.

> **Note**
>
> For a long chain of synchronous calls, it's not a good idea to implement a retry pattern because the consumer will experience a delay if the retry pattern is triggered. We will discuss long chains of synchronous calls in more detail in the *Long chains of synchronous calls – the retry storm* section.

Retry patterns

A retry pattern is implemented to address failures related to service unavailability. It's important to understand different types of failures and only apply a retry policy to transient failures. With a retry pattern, you can retry failed requests based on the retry counter. After every failed request, you increase the retry counter and increase the wait time along with adding some jitter before making another call. Once the retry counter has been reached, you can gracefully handle the failure and let the client know about the unavailability of the service. It's recommended that you increase the wait time between retry requests to allow the service to become available to handle new requests. The intent behind the retry pattern is to provide more execution opportunities to an operation that's likely to succeed. The retry pattern is explained in more detail in *Chapter 4, Keeping the Replatforming of Brownfield Applications Trivial*.

Circuit breaker patterns

Sometimes, a service becomes unavailable for a longer period. In such cases, you don't want to continue retrying the requests as this can result in exhausting your system resources and potentially affecting other services, which will induce cascading failures. Instead, you want to notify your clients relatively early about the unavailability of a service. Circuit breaker patterns help you to address such failures by only allowing requests that are likely to succeed. The idea behind the circuit breaker pattern is that it allows a service to recover automatically by limiting the request load. Once the service becomes available, you can increase the load to serve more requests. You can learn more about circuit breaker patterns in *Chapter 4, Keeping the Replatforming of Brownfield Applications Trivial*.

It's important to understand the behavior of microservices in different circumstances while they communicate over a network. This understanding will help us to increase the availability and responsiveness of the system. In the next section, we will explore the retry storm and how we can address the challenges it presents using different approaches.

Long chains of synchronous calls – the retry storm

When multiple microservices call each other repeatedly in a long chain, there is a possibility that a microservice might take more time to respond than expected, which can result in timeouts. These timeouts will initiate additional retry requests with an expectation that the operation might succeed and the flood of these retries will eventually make the system unusable. This scenario is known as a retry storm, as depicted in the following diagram:

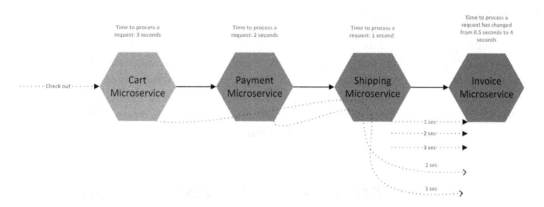

Figure 6.7 – A long chain of synchronous calls

In the preceding diagram, the checkout operation is performed by calling a series of microservices. Each microservice call has its own ETA with a sufficient buffer to address any adverse conditions. Though due to some reason, the invoice microservice is experiencing load and the ETA for the service is no longer the same. The other services are unaware of this change and are expecting the same ETA from the invoice microservice.

Notice how, after 1 second, the shipping microservice has raised a retry request for the invoice microservice and will continue to do so if the response is not received in the due time frame. Similarly, the payment microservice will raise a retry request after 2 seconds for the shipping microservice, which, in turn, raises a request for the invoice microservice. Finally, after 3 seconds, the cart microservice will raise a retry request due to its unmet ETA to the payment microservice, which invokes the shipping microservice and, finally, calls the invoice microservice.

Due to these timeouts, we are raising additional requests without informing other services that its caller is no longer waiting for the earlier response, and they should discard processing those requests. Hence, valuable processing capacity is wasted. Additionally, if you are hosting all your microservices on the same infrastructure, then there is a possibility of cascading failures resulting in the overall system being unavailable.

A distributed transaction (Saga) is an example of calling an ordered sequence of microservices to perform a series of tasks. In the event of failure, retries are performed to ensure consistency across service boundaries. If retries are not successful, the orchestrator needs to perform rollbacks to undo already performed tasks to ensure consistency. The concept is discussed, in detail, in *Chapter 5, Data Design Pitfalls*. There are multiple approaches to how you can handle a retry storm. We will discuss these approaches in the subsequent subsections.

Scalable infrastructure

You can make your infrastructure instantaneously scalable to address peak loads by monitoring different aspects of your application. Instantaneously scalable means that the start-up time of your service is short. This will help us to address issues related to service being overwhelmed by the number of requests. The only exception is if the condition is due to external factors, for example, delays caused due to reading files from a third-party system over a network. In such scenarios, you should look for other alternatives, as we will explain next.

Load balancing, throttling, and request classifications

Load balancing is an approach that is used to distribute incoming requests across a fleet of service instances that are deployed across different servers. These ensure that no server is overburdened while each receives its fair share of requests. Once the server is overloaded, the backend service should reject the requests to inform clients of the right error code or serve a degraded response. Additionally, make sure that you have classified your requests based on their criticality and prioritize them accordingly for execution. In this case, invoice generation is not a critical operation for shipping and can be delayed until the service is restored to full capacity. You can perform rate-limiting at different levels, including the individual microservice, load balancer, or API gateway level.

Exponential backoff, jittering, and circuit breakers

Use the exponential backoff strategy for retry requests. The exponential backoff strategy promotes the idea of exponentially increasing the time between two retry requests. This helps the service to recover by allowing already in-process requests to complete their execution without being overwhelmed by the new ones. Further, you can introduce randomness as part of the backoff strategy to spread out request spikes; this is called jittering. Additionally, you need to ensure that you are not performing endless retries by implementing a circuit breaker and allow the service to recover. Make sure that you have a consistent way of performing retries across your services and avoid retries at different levels, for example, use a service mesh or SDK. You can learn more about evaluating the capabilities of different service meshes in *Chapter 3, Microservices Architecture Pitfalls*.

Only allowing the immediate layer to retry

In this scenario, the shipping microservice is a critical service and only the payment microservice is allowed to perform retries to the shipping microservice. If the shipping microservice has already reached the allowed number of retry attempts, then it should inform the cart microservice that the request can't be served at this time and should be tried later. Therefore, every parent service is informed of the service state. Additionally, it's a good idea to inform the parent/client about the expected time interval in which the service will be available again. In such situations, observing the health of the system is critical; it can help you reject requests early in the life cycle, which provides a better client experience and helps to ensure service recovery in a timely manner by reducing load.

You should now have a solid understanding of the drawbacks of the retry storm, its impact on the overall system, and its potential solutions. We will now discuss event-driven architecture, its pitfalls, and the approaches to overcome them.

The dark side of event-driven microservices

Microservices is a set of loosely coupled microservices implementing bounded context that identifies module boundaries and how they interact with each other. The concept of bounded context is discussed, in detail, in *Chapter 2, Failing to Understand the Role of DDD*. Microservices can be designed as event-driven architecture, where microservices communicate with each other by sharing state changes. The following diagram depicts an event-driven architecture:

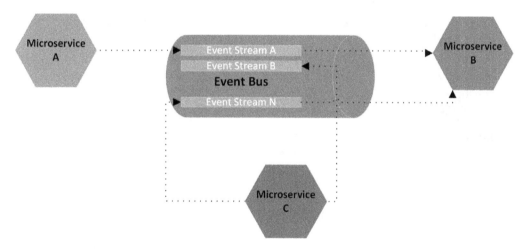

Figure 6.8 – Event-driven microservices

In the preceding diagram, events are produced by producer microservices as an event stream, which is later consumed by multiple consumer microservices. Event streams are served by the event broker (or event bus), which decouples producers from consumers. One of the challenges of event-driven architecture is its difficulty in figuring out how different microservices are communicating with each other via events. One way to address this challenge is to monitor these systems and to track the flow of events and how different systems are reacting to them. In the next section, we will discuss various pitfalls of design events and how they can be addressed using domain events. Let's move ahead and explore them in more detail.

Event design pitfalls

An event is a core concept that enables communication between event-driven microservices. Events promote decoupling by allowing systems to be flexible, resilient, and scalable. All these promises of event-driven architecture depend on how well the events are designed. The following is a list of the different pitfalls that you might face while designing events:

- **Event dependency**: An event that can be affected by past or future events is considered to be a dependent event. In an event-driven architecture, events should be atomic, self-contained, and independent. The consumer should be able to process an event without waiting for other events. Independent events allow different consumers to process the same event differently without relying on the publisher or the publisher making assumptions or intentions about its processing.

- **Entities as events**: Another common anti-pattern is to use database entities as events. These entities are part of the internal implementation of the services and might not covey the right intent in the business context.

- **Generic events**: Avoid using generic events, as this will require consumers to analyze a message before deciding whether they want to process it or discard it, which wastes precious CPU cycles.

- **CQRS, event sourcing, and events: Command and Query Responsibility Segregation (CQRS)** and event sourcing are used together; here, the command and event sourcing provide the write model, while the event store projection and query provide the read model for the system. In such architecture, you'll observe that teams disguise commands as events. Commands are usually for a single consumer, while an event is intended for any consumer that's interested in the event. Another observation you can make is around exposing event store events as a method of communication between different business domains. The downside of this approach is that any changes to an event store are now tied to other business domains, which makes it harder to evolve. Both CQRS and event sourcing were discussed, in detail, in *Chapter 5, Data Design Pitfalls*.

Domain events

In the context of **domain-driven design (DDD)**, domain events are classified as business events. They are considered first-class citizens in DDD, and form the ubiquitous language that's used by stream-aligned teams to understand the business domain. The concept of ubiquitous language was discussed, in detail, in *Chapter 2, Failing to Understand the Role of DDD*. Domain events are specific and contain the necessary information that serves as a contract between the business domains to facilitate communication; for example, the `payment_recieved` event, which belongs to the payment domain, is generated once a payment has been made by the customer. The invoice generation can subscribe to `payment_recieved` and perform the necessary action to generate an invoice. Once an invoice has been generated, the system will generate an `invoice_generated` event, which can be processed by the notification system to send notifications.

An event is a critical part of the event-driven architecture. This is because it acts as a data carrier and a contract between microservices. These events should be domain-driven and represent business events. Both events and API interfaces are ways in which to enable communication. As systems evolve, these interfaces go through several changes, resulting in the release of multiple versions of the service. In the next section, we'll explore the role of versioning in the microservices architecture.

Avoiding service versioning

In *Chapter 1*, *Setting Up Your Mindset for Microservices Endeavor*, we talked about the API-first approach as part of the additional factors of building cloud-native applications in addition to the Twelve-Factor App methodology. The API-first approach emphasizes defining a service contract that dictates the behavior of a service. The service API is a contract between a service and its consumers. The service might undergo many changes during its life cycle, which might require changes to its API. A service could have multiple consumers, and it's crucial to ensure backward compatibility during several API changes to allow different consumers to continue their operations without making any changes.

The API changes can be broadly categorized as breaking and non-breaking. Breaking changes require clients to update to a newer version of the API, while non-breaking changes don't affect clients and allow them to continue working with the older version of the API. Any removal or change in the interface, including the message format, method signature, or response type, is considered a breaking change. The addition of new interfaces, including message formats and message types, are considered to be non-breaking changes.

In the light of the Twelve-Factor App methodology, it is essential to have a single code base for your microservice. This strategy works well with non-breaking changes. However, breaking changes introduce a new challenge where you might have to keep different instances of services running to support different consumers. For example, let's imagine you are decomposing a monolithic e-commerce application into microservices, where these microservices call each other to address different business use cases. A better approach would be to ensure that every change is both backward and forward compatible, which can help in a number of ways, such as allowing rollbacks. In the following diagram, we will discuss the effect of deploying new versions of these services with breaking changes:

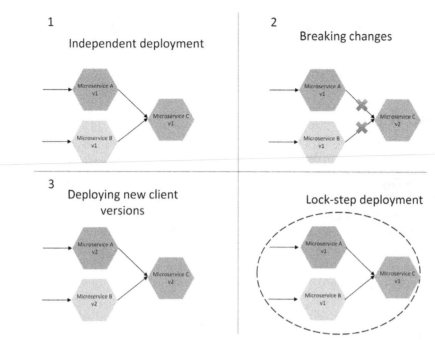

Figure 6.9 – Application versioning

In the preceding diagram, we have three microservices, where Microservice A and Microservice B communicate with Microservice C to get the work done. We can say that both Microservice A and Microservice B are dependent on Microservice C and any breaking change of Microservice C can affect both microservices. The deployment of different microservices and their impact at different stages are discussed as follows:

1. We start with the independent deployment of microservices. All the microservices are running version **v1**. This allows us to deploy new versions of Microservice A and Microservice B independently, but can we do the same for Microservice C?

2. We have a new version of Microservice C; we call it **v2**. The newer version of the microservice has updated the interface; hence, it's a breaking change that affects how Microservice A and Microservice B can invoke the capabilities of Microservice C.

3. Now, both Microservice A and Microservice B have implemented the changes introduced by Microservice C and have deployed their version, **v2**.

4. The team has realized that Microservice C can't be deployed independently; rather, they also need to deploy dependent services. This is what we call lock-step deployment.

Versioning is an important concept that helps us to identify compatibility issues between different components of the system. Avoiding service versioning is a common pitfall that should be addressed early in the development life cycle. In the next section, we will explore semantic versioning and how it works with microservices.

Semantic versioning

A popular approach to address the challenge of service versioning is through semantic versioning. With semantic versioning, you specify a version of a service as **Major.Minor. Patch**. Here, the major number is incremented when you make an incompatible API change, while the minor number is incremented when you add functionality that is backward compatible. Additionally, the patch number is incremented for bug fixes and small fixes. In an SOA or microservices world, incompatible changes should be avoided because of the sheer cost of such a rollout. Teams should avoid such changes even if it means writing extra code. The following diagram depicts the life cycle of a software component and how different versions are created:

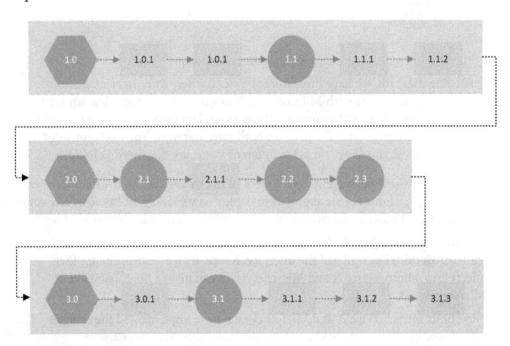

Figure 6.10 – Semantic versioning

In the preceding diagram, a package has been released multiple times with major, minor, and patch updates. Initially, the v1.0 version is released along with two patches (v1.0.1 and v1.0.2). Then, a minor update is released with v1.1. Both the patch update and the minor update are non-breaking changes, thus no changes are required for clients. After a while, a major update has been made to the package, resulting in the release of a new major version, that is, v2.0.

In the next diagram, we will discuss how semantic versioning works with microservices. Microservices are meant to be independently deployable, and it is necessary to allow multiple versions of a microservice with different contracts to run simultaneously. If we don't use semantic versioning, we will get into a trap of lock-step deployment. The following diagram shows how different microservices will continue to operate without any disruption, even when breaking changes are deployed:

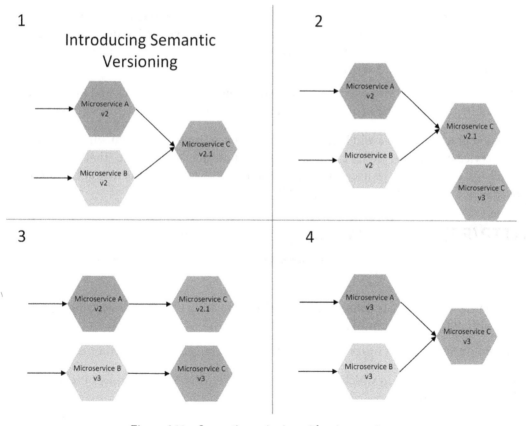

Figure 6.11 – Semantic versioning with microservices

In the preceding diagram, we have introduced semantic versioning to address the challenges of lock-step deployment and how different versions of a microservice can run simultaneously without affecting its clients. Hence, clients can update themselves without causing any disruption or change in deployment cadence. The following is a detailed explanation of the process:

1. A new version (v2.1) of Microservice C is deployed with minor changes, which doesn't require any client updates. In this case, both microservice A and Microservice B will continue to work without requiring any changes.

2. A breaking change is introduced in Microservice C, changing its version from v2.1 to v3. In this case, we will have a new version of Microservice C running side by side to allow Microservice A and Microservice B to revise their communication channel with Microservice C.

3. Microservice B has updated its version to adapt to the new version of Microservice C, while Microservice A continues to work with the older version of Microservice C. Both Microservice A and Microservice B can continue to work and service their clients without any disruption.

4. After some time, Microservice A has also updated its version to v3, and now we can delete the older version of Microservice C.

As a good practice, try to avoid breaking changes. For example, avoid removing attributes from service interfaces. Adding a new attribute is a non-breaking change. That said, the service will need to add the necessary logic to add default values for the new attributes if not provided by the older versions of the API.

HTTP/REST API versioning approaches

Fundamentally, there are two major reasons why we need to update the version of our API:

- Format versioning is defined as a way of manipulating the same underlying resource with different versions of the API.

- Entity versioning is introduced when the underlying resource structure is changed.

There are different ways in which to version your REST APIs. We will explore them in the following subsections.

URI path versioning

With URI path versioning, you can specify the version of an API as part of the URI. However, it violates the principle that one URI should only refer to one resource. When URI path versioning is used with the static binding of a resource, it breaks client integration after every new version release. In the following snippet, the version (v2) is specified as part of the URI:

```
http://api.hrdept.com/v2/employee
```

Query parameter versioning

With query parameter versioning, a query parameter is used to specify the version of the API. Now, every URI represents one resource, and the selection of the right version of the resource can be handled by a single resource handler. In the following snippet, the version (v=1.5) is specified as a query parameter:

```
http://api.netflix.com/catalog/titles/series/123?v=1.5
```

Content-type versioning

The content negotiation versioning technique emphasizes to clients the need to specify the correct version of the API using a custom media type via the Accept header. With this approach, you maintain the same resource URI that's capable of handling multiple versions. The response to this request should contain the version information in a content-type header field. The following snippet depicts the Accept header of an HTTP request that can be used to specify the resource version:

```
Accept: application/vnd.github.1.param+json
```

Custom header versioning

You can use this approach to define a new custom header as part of the request to associate resource versioning. With this approach, you will keep the same URI across different API versions, limiting client code changes between the API versions. The following snippet depicts the custom header of an HTTP request that can be used to specify the resource version:

```
x-eCommerceApp-version: 2021-01-01
```

> **Note**
>
> To learn more about API versioning, please refer to the following link:
>
> ```
> https://github.com/Microsoft/api-guidelines/blob/
> master/Guidelines.md?WT.mc_id=-blog-scottha#12-
> versioning
> ```

gRPC service versioning

The gRPC protocol supports service versioning for both breaking and non-breaking changes. With breaking changes, you should continue to run different versions of your gRPC service side by side unless all of the clients have been migrated to a newer version of your gRPC service. The gRPC package allows you to specify the optional specifier that can be used for versioning, as described in the following table:

Service Address	Package Name		Service Name
ecommerce.v1.notification	ecommerce	v1	Notification
ecommerce.v2.notification	ecommerce	v2	Notification

Table 6.1 – gRPC versioning

The service address is composed of the package name and the service name. The service address allows multiple versions of the gRPC service, which can be hosted side by side to support multiple clients.

Event versioning

In an event-driven architecture, events are considered the core part of the system and are responsible for the communication between different parts of the distributed system. These systems produce, store, and consume events as part of their core structure. Just like any other system, event-driven systems also continue to evolve in how they communicate with other parts of the system by providing a variety of information using an event structure. This requires us to think carefully about publishing events that can be processed by a variety of consumers without making any changes to the consumer or, at the very least, provide enough time for consumers to update the implementations to adhere to the new event structure.

As a rule of thumb, a new version of the event should be convertible from the old version; otherwise, it's a new event. It's a good practice to specify the event version as part of the event structure. This is because it gives you the flexibility to publish multiple versions of an event to allow different consumers to process it. Another approach is to go with no event versioning and keep the events backward compatible by ensuring that you don't delete or change fields in the event structure. New fields are treated as nullable values so that the old version of the events can be replayed later. The following diagram depicts how event versioning works in an event-driven architecture:

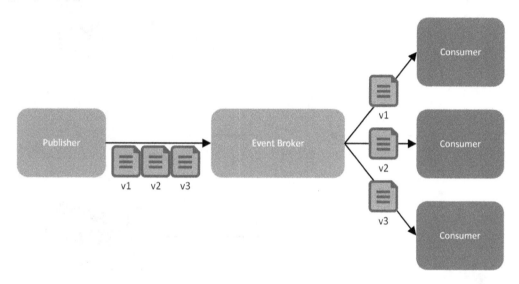

Figure 6.12 – Event versioning in an event-driven architecture

In the preceding diagram, the publisher is producing multiple versions of an event to allow multiple consumers to react to those events. Ideally, these consumers should be idempotent to avoid the side effects of processing multiple versions of the same event. The event broker acts as a relay between the publisher and the consumer and provides capabilities of filtering events based on event data. The event broker helps decouple publishers and consumers. The idea is to allow different parts of the system to evolve independently, where events aid in enabling communication, where each consumer processes relevant events.

In the event sourcing system, the system should be able to recreate the application state using events created during the life cycle of the event source system. This implies that the system should process both old and new events to recreate the application state. The following diagram depicts multiple versions of the different events in the event store. Later, we will discuss the challenges associated with this approach and how you can address them:

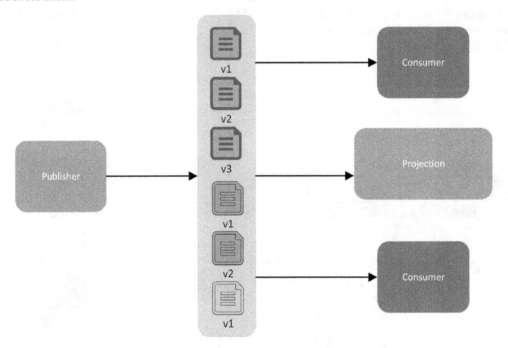

Figure 6.13 – Multiple versions of events in the event store

In the preceding diagram, the event source system has multiple versions of the same event in the event store, which are used for projections, and downstream systems for consumption. The challenge with multiple versions is that the system needs to select the right version while recreating the state, which makes the overall system complicated and error-prone. The two approaches, how event source system process events are, **strong schema** and **weak schema**. A strong schema is based on a type system, where the system can only deserialize events to types that exactly match their schema. Any new events require building support for serialization, which results in redeployment.

On the other hand, weak schemas are more adaptive and can be described in JSON or XML format. Weak schemas use mapping techniques to map JSON or XML data to event objects; this is as long as we don't change the field names and their semantic meaning. A weak schema helps us to keep the latest version of the event, as old events are now convertible to the latest event. A weak schema is the recommended approach for handling event source systems. Other approaches include versioning the event store by replacing the old events with the new events while keeping backups of the old event store for auditing.

With a good understanding of the different versioning techniques that can be applied in a microservices architecture, you can make sure that the system will continue to operate without any disruption even when breaking changes are deployed. In the next section, we will explore the role of service meshes in simplifying and scaling inter-process communication between microservices.

Service meshes

A service mesh provides an abstraction layer on top of microservices to facilitate inter-process communication. The idea behind a service mesh is to decouple communication concerns from the implementation of microservices and handle them in a consistent manner across the microservices architecture. The following diagram depicts a simple service mesh architecture for microservices:

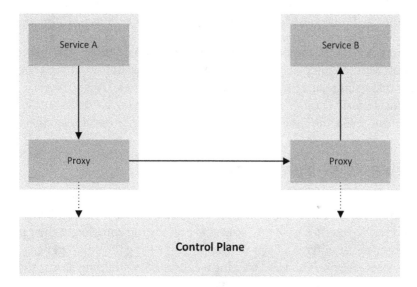

Figure 6.14 – A service mesh

The service mesh is a configurable communication infrastructure layer that guarantees the reliable delivery of requests across a complex topology of services. This enables teams to focus on business capabilities and leave the communication concerns to the service mesh. The service mesh is composed of two components: the control plane the and data plane. The data plane is implemented as a sidecar proxy that's responsible for intercepting data packets across the network and performing various functions before forwarding. The data plane also performs routing, health checks, load balancing, observability, authentication/authorization, and service discovery. On the other hand, the control plane provides management and monitoring concerns. It acts as a controller that connects sidecar proxies to form a distributed system.

In a service mesh architecture, the microservices are unaware of the distributed system, as they only talk to their local sidecar proxy for communication with other microservices. The service mesh technology is maturing as more and more organizations are adopting it. With the adoption of the service mesh, there are a few pitfalls that you should be aware of:

- The management of service meshes should be part of shared governance, where the enablement team should work with value stream teams to ensure that they understand the benefits and challenges rather than implementing different communication concerns that can be handled holistically by the service mesh.

- Like any other technology, a service mesh is not a silver bullet, and it requires a steep learning curve. The deployment and operational cost of a service mesh are high. Earlier in the book, we compared different service meshes based on their capabilities in. Please refer to *Chapter 3, Microservices Architecture Pitfalls*.

The service mesh is a good candidate if you are having trouble scaling inter-process communication between microservices. It manages the communication between microservices that helps teams to focus on business capabilities. Usually, a service mesh is owned by the platform team. In the next section, we will compare Dapr and service meshes to help you make the right choice while building a microservice architecture.

Dapr versus service meshes

Both Dapr and the service mesh offer overlapping capabilities, raising the question of how Dapr compares to a service mesh and when one should be used over the other. While both Dapr and service meshes have the same goal of reducing complexity, their intended audiences are somewhat different. Dapr focuses on improving the developer experience by abstracting your microservices and offering a collection of building blocks to meet common microservices needs, whereas a service mesh helps application operators by providing a network service mesh that addresses networking concerns. Let's take a closer look at the similarities and differences between the two, as listed in the following table:

Features	Dapr	Service Mesh
Service-to-service invocation	Yes	No
State management	Yes	No
Publish/subscribe	Yes	No
Resource binding and triggers	Yes	No
Secrets	Yes	No
Actors	Yes	No
mTLS	Yes	Yes
Metrics	Yes	Yes
Resiliency	Yes	Yes
Distributed Tracing	Yes	Yes
Traffic routing	No	Yes
Traffic splitting	No	Yes

Table 6.2 – Comparing Dapr and service meshes

So, should you employ Dapr, a service mesh, or a hybrid of the two? Well, the answer depends on your needs. When using both Dapr and a service mesh, it might seem intuitive to activate the common capabilities (**mTLS**, **Metrics**, **Resiliency**, and **Distributed Tracing**) of either one of them, but this depends on your needs. In a hybrid approach, it is advised that only Dapr or the service mesh be configured to execute mTLS encryption and distributed tracing.

For distributed tracing, a service mesh only supports synchronous communication, while in the case of communication via a pub/sub mechanism, the calls are not traced. Dapr supports tracing by embedding trace IDs into cloud event envelopes, which enables tracing for both synchronous and asynchronous communication. Hence, Dapr is a better option since it offers end-to-end tracing for a microservice architecture with both synchronous and event-driven communication, whereas a service mesh is a suitable fit if the design only supports synchronous communication.

In terms of resiliency, not all service meshes are equal. For example, a circuit breaker is not available on Linkerd. On the other hand, a retry policy is supported by both service meshes and Dapr. Although there is a significant difference between the two, where a service mesh allows you to define retry policies per service, Dapr allows you to specify retry policies per call.

Summary

In this chapter, we learned about various communication styles that can be adapted to facilitate communication in a microservices architecture, their challenges, and potential solutions. We discussed the fundamentals of system communication and how different architecture styles use them for communication. Further, we analyzed the drawbacks of direct client-to-microservices communication and how we can address them by introducing an API gateway that acts as a reverse proxy. We also explored how using a single API gateway for large applications can create ambiguities and how we can address these complexities using different approaches.

Later, we discussed availability and responsiveness as two important concepts and how different techniques can be used to enable resilient inter-process communication between microservices. Additionally, we looked at how a long chain of synchronous calls can lead to a retry storm that can make the overall system unavailable along with multiple approaches that can ensure availability and performance at different levels. We also discussed event-driven architecture and its challenges and how we can address them using different approaches. We also discussed the role of versioning in different aspects of microservices. Finally, we explored the evolution of a service mesh and how it compares to Dapr.

With this knowledge, you can start your journey of selecting the right communication mechanisms to enable client-to-service and service-to-service communication using the approaches presented in this chapter. In the next chapter, we will discuss the pitfalls related to cross-cutting concerns and how they can be addressed using different solutions.

Questions

1. What's the difference between synchronous communication and asynchronous communication?

2. What are the disadvantages of direct client-to-microservice communication?

3. What is a retry storm?

4. What are the different approaches to HTTP/REST API versioning?

5. What is a service mesh?

Further reading

- *Hands-On RESTful API Design Patterns and Best Practices*, *Packt Publishing*, Harihara Subramanian and Pethuru Raj

- *Enterprise API Management*, *Packt Publishing*, Luis Weir

- *Building Event-Driven Microservices*, *O'Reilly Media*, Adam Bellemare

- *Mastering Service Mesh*, *Packt Publishing*, Anjali Khatri and Vikram Khatri

- *Microservice Patterns and Best Practices*, *Packt Publishing*, Vinicius Feitosa Pacheco

- *Versioning in an Event-Sourced System*, Gregory Young

7
Cross-Cutting Concerns

Traditionally, when we start building applications, we develop components known as *facades* to perform different operations, such as logging, health checks, tracing, and authorization. In microservices architecture, building these components for each service requires proper time and effort. Microservices are self-contained services that generally require less time to build because they are limited to a single business capability. Building the modules for cross-cutting concerns for each service would certainly take a much longer time than building that service itself. As a consequence, it affects agility and requires adequate effort to develop. Since each service has its own components, modifying one component due to an organizational plan would entail updating each service individually, which would be challenging as well.

It is a good idea to build common modules or reuse existing platforms that allow easy and fast integration with different microservices. Microservices should only include modules or artifacts specific to the business capability, and all cross-cutting concerns should be outsourced to solve the problem. For example, instead of building a separate logging mechanism for each microservice, it is preferable to use a centralized monitoring system such as **Azure Application Insights**, which provides an **SDK** and easy integration with whatever platform the microservice is based on. Furthermore, an Azure App Insights resource can be integrated with all microservices and provides unified access to all the logs from a single pane of glass. However, when you are keeping one resource for the whole system, it becomes harder to restrict access due to privacy or legal needs. In such a scenario, you can segregate log resources based on services or domains.

In this chapter, we will discuss the pitfalls related to cross-cutting concerns, and also learn patterns such as microservices chassis, gatekeeper, and service discovery. Throughout the chapter, we will cover the following topics:

- Microservices chassis patterns
- Cross-cutting pitfalls

Microservices chassis patterns

In the non-technical term, we can think of a *chassis* as the base frame of a car. It provides the core layer, or skeleton, on which the car is built. In the context of microservices, we can think of it as a base framework that provides basic cross-cutting concerns such as logging, security, exception handling, tracing and distribution, and more. The following figure depicts the representation of the microservices chassis framework:

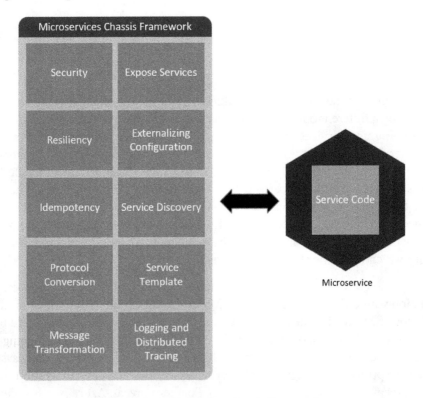

Figure 7.1 – Microservices chassis framework

Microservices architecture for large enterprises has numerous services communicating with each other, and agility is a very important factor. When the requirement comes, you usually estimate it based on the business use case instead of thinking about how to implement security, develop a logging framework, or perform exception handling, and so on. Providing a base framework that can be utilized by the service to handle these concerns not only reduces the number of efforts required to build that service but also allows you to reuse those building blocks that are part of the chassis framework.

The microservices chassis pattern states that all of these concerns should be offloaded to either a distinct service that is shared by all other services or a component. **Dapr** is a good example of a service that provides a set of building blocks to enable logging, service-to-service communication, inter-process communication, pub/sub, and a few others. It also requires very little configuration and allows quick implementation by using the HTTP or GRPC endpoints, or through Dapr SDK. To learn more about Dapr, refer to *Chapter 3, Microservices Architecture Pitfalls*.

Microservices application needs

There are various scenarios and use cases built on microservices. Microservices are not only dependent on the business capability or **Create, Read, Update, and Delete (CRUD)** operations performed in the database, but can also be built to respond to events that arrive at the message queue, integrate with other systems, be exposed as a webhook to send some notification, or perform some heavy operational task, such as image processing. For each service, you may need to build several components that are not completely tied to the actual service use case and would delay the service delivery. Some of the factors that should be catered for separately from the microservice but not limited to are as follows:

- Security and token validation
- Retry and circuit breakers
- Idempotency
- Protocol conversion
- Message transformation
- Logging
- Metrics collection
- Distributed tracing

Building all these concerns in the microservice itself is an anti-practice and increases the development time of the service, therefore reducing agility. In the next section, we will discuss some of the pitfalls when catering to these concerns and the recommended practices for dealing with them.

Cross-cutting pitfalls

This section will cover some of the pitfalls related to various cross-cutting concerns when building microservices, followed by an alternative approach for each.

Embedding security handling inside microservices

Protecting services and allowing secure access to services are some of the primary needs when building services. Embedding the security process for validating an incoming token from a request object into the microservices will be a time-consuming task. Security and token validation can be eradicated from the microservices code, keeping the microservices code clean and tied to its use case. However, if the service does not validate the incoming token, make sure the service is not exposed directly to the consumer without having a front door such as an API gateway. If the service is exposed, it will be vulnerable to attacks and may fail. The solution to this pitfall is by offloading the security to the API gateway.

Enable throttling to protect microservices

Throttling is a mechanism to limit the number of requests allowed for a component. In a microservices architecture, it is considered to be an important factor when it comes to protecting your services. If you don't have a mechanism to configure the maximum number of requests your service can serve, it may lead to a **Distributed Denial of Service (DDoS)** attack. Throttling allows you to specify limits, configurations, and policies, which enables services to meet their expected **Service Level Agreements (SLAs)**.

> **Note**
> The SLA is a commitment between a service provider and a customer that identifies both the services required and the desired level of service.

Offload security token validation to API gateway

There are various services in Azure where **Azure API Management** provides a **JWT** policy to validate the incoming token and route it to the backend microservice. You can set up Azure API Management with microservices hosted inside the **Azure Kubernetes Services (AKS)** cluster that allows the external traffic to reach Azure API Management and route it to the backend service inside a **VNET**. The following figure illustrates using Azure API Management in conjunction with the microservices hosted inside the AKS cluster with VNET configuration:

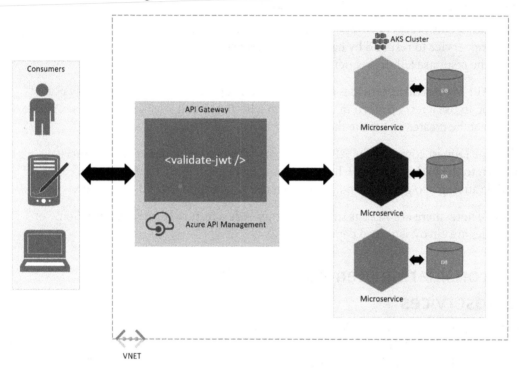

Figure 7.2 – Validating token at API gateway level

In the preceding figure, the AKS is configured with a VNET configuration and the microservice is only available to the services within the VNET. Azure API Management is configured with an external VNET configuration that allows all the traffic from the public internet (that only reaches to the service) inside a VNET.

Not considering resiliency when building microservices

Developers who are building APIs for enterprise-grade applications may understand the need for retry and circuit-breaker patterns. We have discussed these patterns in detail in *Chapter 4, Keeping Replatforming of Brownfield Applications Trivial.* The solution to this pitfall is by using retry and circuit breaker patterns.

Using retry and circuit-breaker patterns

Retry and circuit-breaker patterns must be applied to create a robust mechanism, allowing the failing service to respond by having several retries or opening the circuit for a short time if the response failure frequency is high.

Azure API Management provides an out-of-the-box option to configure the retry policy. However, to use a retry pattern in conjunction with the circuit-breaker pattern, a separate service can be created to handle these concerns.

When the request comes to the API gateway, it will be routed first to the retry service and then to the backend service. If the backend service is failing, the retry service makes multiple attempts to get the response or open the circuit for a limited timeframe.

As a side note, there are various sidecars that provide a feature of retry pattern. Some of the names are envoy, istio and dapr.

Not considering idempotency when building microservices

In microservices, the system is decomposed into several services, listening over HTTP endpoints. Today, **RESTful** services are an industry standard for building APIs. Since RESTful services rely on the underlying HTTP protocol, they use various HTTP verbs for different operations. For example, the HTTP GET method is used to retrieve the data, the HTTP POST method is used to write, HTTP PUT for update, and HTTP DELETE for deletion. Since the HTTP POST method writes data on a server, it is not safe. If the same operation is executed twice, multiple entries will be created on the server. To protect services from these issues, idempotency is an important factor to consider.

> **Note**
> Binary protocols such as **gRPC** and **thrift** are fast overtaking REST as the preferred way to build new services.

Coming from a monolithic perspective, where the application is typically divided into layers and communication between those layers is achieved via direct references, ignoring idempotency may work. With a microservices architecture, the architecture should be designed in such a way that if any service fails and the same operation is repeated, the data should not change.

The following table shows the list of common HTTP verbs with their idempotency and safety states:

Operation	Idempotent	Safe
GET	Yes	Yes
POST	No	No
DELETE	Yes	No
PUT	Yes	No
PATCH	Yes	No
HEAD	Yes	Yes
OPTIONS	Yes	Yes
TRACE	Yes	Yes

Table 7.1 – HTTP verbs idempotency and safe values

In the preceding table, GET, DELETE, and PUT are idempotent methods since if the same record is retrieved, deleted, or updated, the data will not change. However, it is not safe as it executes the same operation on the server again. The solution is to implement idempotency in microservices architecture.

Implementing idempotency using keys

There are several ways to achieve idempotency; however, the most common way is to assign a unique key to every transaction being executed. The generation and assignment of the idempotency key should be considered a cross-cutting concern that is commonly implemented as part of the API gateway. Alternatively, this responsibility can be delegated to a common service, solely responsible for addressing cross-cutting concerns to keep the API gateway implementation simple. The key can be a **GUID**, or any random number generated that gives an identity to the transaction.

On the service's database side, we can keep this key for each table, thus avoiding creating or modifying the record if the same key is present. So, if any transaction fails due to service unavailability or any other reason, the same transaction can be processed and repeated, and protected from creating multiple entries in the database if the same transaction key is present. The following figure depicts the use case of attaching the idempotency key at the gateway level and keeping that in the service's database to avoid repetition:

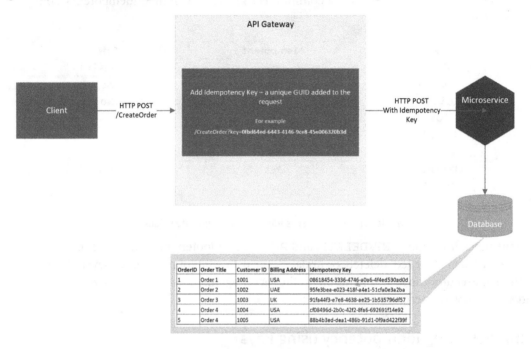

Figure 7.3 – Adding idempotency key at API gateway level

Idempotency is a very important design principle for building web API-oriented architecture, such as microservices, or service-oriented architectures, and helps avoid data inconsistencies and duplication.

> **Note**
> Gateways usually never open up as much of the request to know where to route the request. Adding such a core responsibility to the gateway is going to cause serious issues in the coming years or decades.

Embedding protocol translation inside microservices

Usually, the microservices' core APIs are independent, fine-grained services, and exchange data over HTTP. For example, when integrating a microservices application with an existing **SAP** system, you may need to use the TCP protocol to send data. To meet a specific requirement, you can create components or services that allow other services to communicate with them using the HTTP protocol and connect to them internally using a TCP protocol. The solution to this pitfall is to delegate the protocol translation responsibility to the API gateway.

Delegating protocol translation to API gateway

Building the protocol translation inside the microservice itself is an anti-pattern. You should delegate this responsibility to either an API gateway, or, build a common service to perform this translation which makes the service generic, and then the same translator can be used by other services as well.

Protocol translation allows developers to concentrate on programming instead of how to access the microservices. Rather than writing code to execute protocol translations, developers can focus on designing microservices to cover specific business scenarios and delegate this responsibility to the API gateway. For example, making REST calls from the **Single-Page Application (SPA)** is far easier and more lightweight than the **SOAP** protocol, which only works with XML messages. Clients can consume services over REST APIs, and the gateway can take responsibility for all the translation needed and consume the target service over the SOAP protocol.

The following figure depicts the use case of doing protocol translation at the API gateway level:

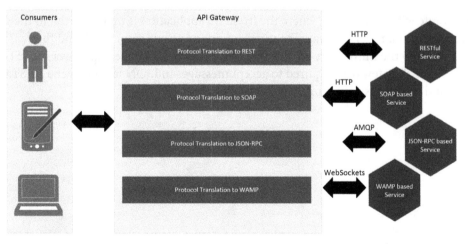

Figure 7.4 – Protocol translation at the API gateway

The API gateway helps developers to offload the protocol translation, and it can become a common entity that holds all kinds of translation components that help translate protocols. All requests from clients first go through the API gateway and then route to the appropriate microservice.

The API gateway is often used to handle requests by invoking multiple microservices and aggregating the results. It allows you to translate between web protocols such as HTTP and **WebSocket** and web-unfriendly protocols that are used internally.

Message transformation coupling with microservices

REST services are one of the primary architecture styles used today when building services or APIs. However, there could be scenarios where you are decomposing a monolith into the microservices application, and you want to integrate with some legacy services to complete a specific functionality, for example, transforming messages into different formats so they can be accepted by the target service. The solution to this pitfall is to either delegate this responsibility to the API gateway to transform the messages or build a separate service.

Building a separate service or using the API gateway to provide message transformation

Keeping messaging transformation within the microservice itself is an anti-practice. The drawback of embedding this in the microservice is it reduces reusability and violates the SRP. The recommended way is to offload the message transformation to the API gateway level. If any of the services want to transform a message, they can communicate to the API gateway to perform that work.

Let's elaborate with an example, where the frontend application is making a call to the microservice over an API gateway. The microservice performs some work and then communicates to the third-party service that only accepts SOAP messages over the HTTP protocol. The API gateway is required to accept messages in JSON format over a REST standard and then transform it into the SOAP envelope:

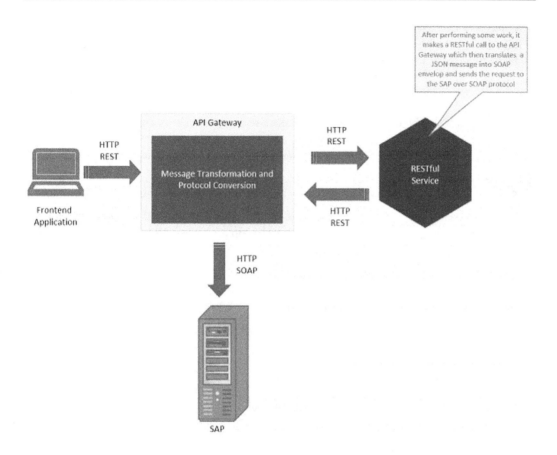

Figure 7.5 – Message transformation and protocol translation and API gateway level

Azure API Management provides an out-of-the-box feature to expose your SOAP-based services to RESTful endpoints. When adding the API, you can select the **WSDL** definition template to configure SOAP-based APIs, as shown in the following figure:

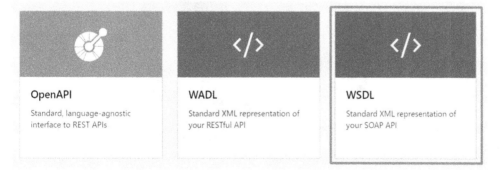

Figure 7.6 – Message transformation and protocol translation at API gateway level

More importantly, Azure API Management is one of the most popular services in Azure when it comes to implementing microservices chassis patterns. It helps embed various policies at the gateway level to protect and control your backend services.

> **Note**
>
> To learn more about Azure API Management, refer to `https://docs.` `microsoft.com/en-us/azure/api-management`.

Directly exposing microservices to consumers

Microservices-based applications are split into a number of fine-grained services where each service is hosted in its own container and listens on a different IP address or port. The client application connects to certain services to do some work, but exposing the service's IP address directly to client applications or other consumers is a bad practice. If the IP address of a service changes due to scaling, failures, or upgrades, the client code needs to be updated, which can cause a delay and could have a negative impact on the business.

The following figure depicts the approach of client applications accessing services directly:

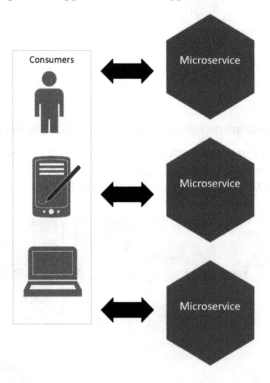

Figure 7.7 – Microservices exposed directly to consumers

The drawbacks of exposing services directly to consumers are as follows:

- Consumer application needs to track all the services' endpoints. With any change in endpoints, consumer application needs to be updated.

- It establishes coupling between a service and a consumer application.

- All the RESTful methods are exposed to consumers regardless of whether that is needed or not.

- Each service needs to handle the authentication and authorization of requests.

- Services with public endpoints are a possible attack surface that must be minimized.

With all of these disadvantages, implementing a *gatekeeper* pattern is a preferred solution to resolving this problem.

Using the gatekeeper pattern

Gatekeeper patterns protect services by providing a dedicated host instance that acts as a broker between consumers and services. The gatekeeper helps to validate and sanitize the request and routes the request to the backend services. It can be used in the following additional scenarios:

- When security is your utmost concern, and you need to protect your backend services by enforcing restrictions and policies

- To validate all the incoming requests before routing them to the backend services

- To expose only those APIs that need to be exposed

The following figure depicts the use case of the gatekeeper pattern that acts as a piece of security middleware between frontend and backend applications to protect them:

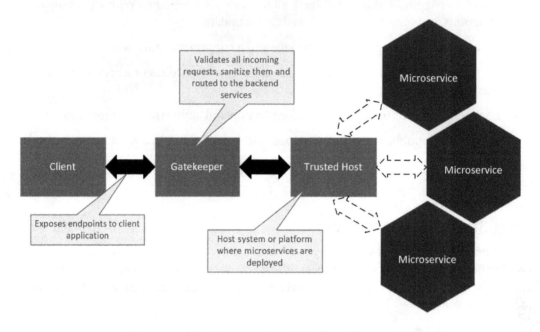

Figure 7.8 – Using the gatekeeper pattern

The preceding figure shows the use of the gatekeeper pattern that validates and sanitizes requests before sending them out to the backend services. Gatekeeper patterns can be implemented by adding the API gateway as a proxy to backend services. The **Trusted Host** is the system or platform used to host the microservices. Kubernetes could be an example of a **Trusted Host** in microservices architecture.

A few considerations to understand regarding gatekeeper patterns are as follows:

- Gatekeepers should not hold any tokens or access key information.

- Gatekeepers should not contain any logic to perform operations related to services or data.

- Gatekeepers should be designed as a single point of failure and there should not be any room to access the services directly.

- You should consider implementing monitoring and logging at the gateway level.

- Always use secure connections when communicating to or from frontend or backend services and the gatekeeper. Use secure connections such as HTTPS, SSL, and TLS.

The gatekeeper pattern is one of the most common patterns used when building enterprise-grade applications. In Azure, **API Gateway**, **Azure Front Door**, and **Azure** API Management can be used to implement a gatekeeper pattern.

Keeping configuration inside microservices

Every business application needs to connect with various systems to perform business operations. To communicate with systems that could be databases, message brokers, services, or other platforms, you need to provide a key or a connection string to access them. Keeping all of these configuration values in the microservice itself is an easy option, but it introduces tight coupling when you want to connect to other environments. The solution to this pitfall is to externalize your configuration values, so instead of changing the values at the application level, when deploying to that environment, you can switch these values from the external system to the target respective environment.

Externalizing configuration

Externalizing configuration is the practice of decoupling the configuration values from the application. One approach is to set the configuration values as parameterized values when running the application and then building the logic inside the application to read those values at runtime. On the other hand, you can create a data store that holds all of the property values, and the application can read them at runtime depending on the context in which it is functioning. The application does not need to be recompiled with this approach, and the values can be conveniently maintained.

When using DevOps for automating build and release cycles, configuration values can be defined in the pool of variables collection, also known as *variables groups* in the context of Azure DevOps. The values can easily be changed and used with respect to the environment you are targeting.

Using Azure Key Vault

Azure Key Vault is a cloud-based service for storing secrets. You can keep all the configuration values or properties inside Azure Key Vault and use it as an external configuration storage provider. All the keys/values stored in this resource are highly secured and it enforces the TLS protocol to protect them between clients and Azure Key Vault.

You can store all your application's secrets at one central location that is closely regulated and secured, and you can access them safely by using their special **URI**s. Furthermore, proper authentication and authorization are required for any access. Authentication can be accomplished in a variety of ways, including **Azure Active Directory** authentication, **Role-Based Access Control** (RBAC), or through a Key Vault access policy.

> **Note**
> To learn more about Azure Key Vault, refer to `https://docs.` `microsoft.com/en-us/azure/key-vault/general/basic-` `concepts`.

Some of the best practices when using Azure Key Vault are as follows:

- Create access policies for Azure Key Vault.
- Enable least privilege access to Azure Key Vault.
- Turn on firewall.
- Enable VNET service endpoints if the other cloud resources are accessible over VNET.
- Enable soft delete to protect your keys from accidental deletion.
- Set up automation to automatically respond when your secrets rotate.
- Store all security certificates in your vault.
- Ensure that the deleted key vaults are recoverable by enabling the built-in protection features of Azure Key Vault.
- Enable monitoring and set up some metrics to monitor the vault's availability, saturation, latency, and error codes.
- Set up alerts to send out notifications when any of the metric conditions suffice.

Undiscoverable microservices

In a microservices architecture, services are hosted separately on a different IP and port. Unlike a monolithic application, where you can add a direct reference to the module or component you want to use, with microservices you need to make an HTTP/REST call to invoke the operation of the target service. Since the system with microservices is divided into sets of various services, you need to build some mechanism to make your services discoverable.

Services are either consumed by the client or frontend applications or by other services to perform service-to-service invocation. To do so, implementing service discovery patterns at both client and server levels is important. To clarify our understanding, in the next section, we will discuss the client-side service discovery and server-side discovery patterns, with examples.

Implementing client-side service discovery patterns

Microservices usually run inside a containerized environment, where multiple instances of a single service can run in parallel and each instance is assigned a unique IP address or port. The problem is – how does the client know if the IP changes or is dynamically assigned?

The solution to this is by implementing the *service registry* pattern. A service registry registers all the services running across the cluster and lets clients request the service endpoint by providing the service name or key. A service registry can be considered a database of services, their instances, and other information, such as IP addresses and ports. The client application searches the service registry to obtain the list of available running instances and selects the one that best serves the load balancing criteria. The `heartbeat` function is used by the service registry to check the availability of services.

The client must be able to access the service registry and you must write some code to connect to the service registry. After getting the details of the active instances, the client must decide to use the correct endpoint to connect to that microservice directly. This kind of client is also known as a *registry-aware client*.

The following diagram depicts the use case of a service registry in a microservices architecture:

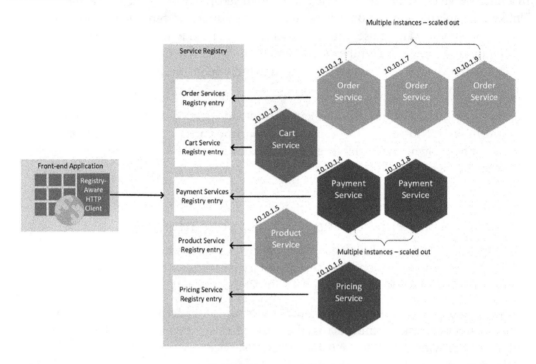

Figure 7.9 – Client-side service discovery pattern

In the previous figure, the frontend application is a registry-aware client application that communicates to the service registry over HTTP and gets the list of running instances details to access them directly. On the other hand, microservices should be capable of self-registering to the service registry when they start and unregistering when they shut down. Moreover, any third-party registration service can also be used to register/unregister services to the service registry.

Implementing server-side service discovery patterns

With a server-side discovery pattern, the client can make a request to the load balancer that acts as a front door to the backend services and route the traffic based on the load balancing algorithm. In this scenario, there is no service registry where the client can request the available endpoints. Instead, it contains a load balancer endpoint that can be used to make a request.

In the AKS world, the concept of a Service type object with a load balancer configuration is the example used for server-side discovery patterns. The Service type object exposes the external IP address that can be consumed by any client application and internally it maps the request to the respective pods where the actual containers are running.

The following figure depicts the architecture with a server-side discovery pattern:

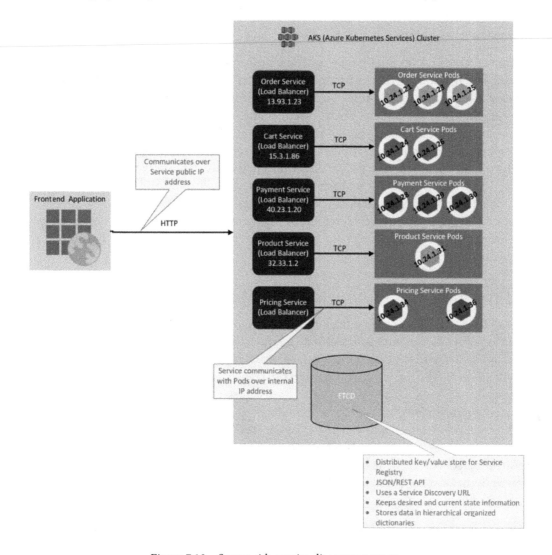

Figure 7.10 – Server-side service discovery pattern

The preceding diagram depicts the implementation of a server-side discovery pattern inside an AKS cluster. It shows the microservices containers running inside pods, where each pod has a unique IP address assigned within the Kubernetes cluster. Since the pod's IP address is not exposed to the outside world, the K8s (Kubernetes) service's object is created for each set of pods to help load balance the requests and establishes the mapping between services and pods. The configuration, and any other desired and current state information, is stored in the ETCD database, which is the service registry in K8s to store information in a key/value format.

The frontend application can directly consume the services over a public IP address without writing a code to get the list of running instances and then building a logic to consume the available endpoint, as we saw in the client-side discovery pattern. With this approach, the client applications are not aware of the service registry as it is maintained on the server-side. Moreover, you can also handle third-party clients easily, and less maintenance is required. Apart from exposing services such as load balancers, there are a variety of other methods, such as **NGINX**, which can be used as a server-side discovery load balancer. This can take all requests from the client application and route them to the K8s service object, configured as Cluster IP, which then further directs traffic to the corresponding pods where microservices containers are running. Service meshes usually solve the service discovery problem for service-to-service invocation and API gateway-to-service calls.

Not maintaining a service template or framework

Many software companies maintain a practice to build in-house frameworks that can be reused to reduce development efforts. With a microservices architecture, the application is decomposed into several services, and each service is built by a separate team or set of developers. The service is aligned with the business capabilities, and assigning responsibility for building components to address cross-cutting concerns for each microservice's team is an anti-practice, and not considered a viable solution. Each team needs to spend some time, which may be days or weeks, to build those components that can be used to perform logging, event messaging, connectivity with databases, and other cross-cutting concerns. An alternative solution to this pitfall is building a template or a framework that can be reused by all the services to reduce this effort.

Building a service template or framework for quick development

When building a service template for quick development, you should make sure that it is platform and technology-agnostic. Many companies reuse existing code bases while building new apps. Companies develop in-house frameworks, for example, in **.NET**, and reuse them as a basic template for developing new projects, or by adding a direct reference to that project to leverage the existing code base's capabilities. With microservices, the technology could be different for each service, so sharing the same code base would not work. For example, if a service project is based on **.NET Core**, we cannot add it as a reference into a **Node.js** project.

An approach that can be used is to build a framework that allows other services to interact over standard interfaces or protocols such as HTTP or GRPC, or by building an SDK that can be used by the microservice's team. At the start, you can think about interacting with this layer over HTTP or GRPC endpoints, and further invest time to build SDKs for different platforms. This framework should be built in a way so that it does not hold any microservice's specific business logic. It should be independent and provide standard or generic interfaces to allow integration with any service or platform. One good example of this is the Dapr framework.

Some of the benefits of building a framework are as follows:

- Developers can build microservices focusing on the business domain instead of developing components for cross-cutting concerns.
- Provides a centralized place to update or introduce new components that can be leveraged by many services.

Some of the drawbacks include the following:

- Separate time and effort are needed to build the framework.
- Expertise is needed to build SDKs for each language or platform.

Not paying attention to logging and monitoring

Implementing logging is very important with microservices applications and it helps to troubleshoot issues quickly and easily. Sometimes, a single transaction spans multiple services, so keeping each service's log separate would make it difficult to investigate or identify the root cause of an issue. Therefore, it is recommended to implement centralized logging for the application, and each service should write the logs to that centralized store.

Azure Application Insights is one of the prevalent services in Azure that allows easy integration and extended monitoring capabilities. Either utilizing the ready-made framework or a service that allows a quick adoption and implementation of logging into your services is what addresses the attributes of the microservices chassis pattern.

Logs, metrics, and traces

For monitoring microservices logs, metrics and traces are the core elements. Each plays an important role when it comes to application monitoring. Let's explore each of them:

- **Logs**: Every application is required to log information that can be used to watch how an application behaves or to troubleshoot errors. Logs should be captured with severity levels that allow developers or administrators to easily filter and take appropriate action when needed. With logs, you can keep as much information as possible and assign some parameters, such as a unique number or date-time information.

- **Metrics**: Metrics include aggregated results, and usually contain a name, timestamp, and a field to represent a metric value. Metrics are used to gauge the health of the application and whether it is running or stopped, but, in order to get detailed information, you need to access logs.

- **Traces**: Traces contain information about specific application operations. Traces provide complete visibility of the health of an application. Traces provide less information on the infrastructure or the environment where your application is running, and for that, you need to check the metrics.

Implementing centralized monitoring

When implementing monitoring for a microservices-based application, it is better to implement it as a centralized monitoring mechanism. Keeping the monitoring separate from each service increases agility in finding the issues, and also gives each team the autonomy to use whatever technology they can for monitoring purposes. Therefore, centralized monitoring reduces the need for understanding different monitoring technologies implemented by different services, and also helps to filter out the information swiftly to solve a problem.

Centralized monitoring helps teams to log everything in one central place and it also helps to identify the issues quickly and easily. Consider a scenario where a customer wanted to purchase an item through a shopping cart system. When a customer checkouts the items in the cart, it first proceeds to the payment gateway through the payment service, which also saves the payment information. Once the payment is processed, the order is generated, and the notification will be sent.

Now, in this scenario, there are four services involved, namely Cart Service, Payment Service, Order Service, and Notification Service. If you notice that a single transaction is spanning over four services, and if any of the transactions fail, you need to look up the logs to investigate or identify the issue. If the logs are maintained separately, this would be very difficult and time-consuming. However, with centralized logging, the user can easily query the monitoring platform to find the issue and get a complete understanding of when the request was raised and where it failed.

Azure Application Insights is one of the most widely used resources in the industry for monitoring applications. It sends telemetry from the application to the portal that can be analyzed to determine the performance and usage of the application. The Azure Application Insights telemetry model is standardized, making it easy to implement monitoring that is platform and language-agnostic. The following figure depicts the scenario of a shopping cart system that is using Azure Application Insights as a centralized monitoring platform:

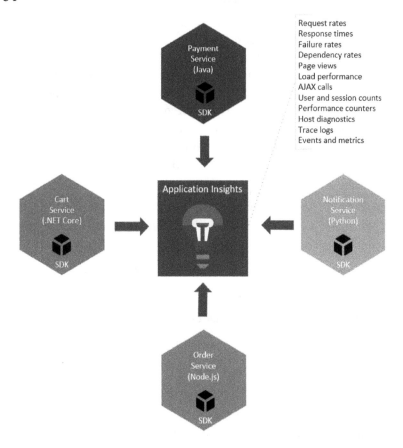

Figure 7.11 – Azure Application Insights for centralized monitoring

The preceding figure illustrates the concept of having Azure Application Insights as a centralized monitoring system, and each microservice is communicating to that service using the Azure Application Insights SDK.

> **Note**
>
> To learn more about Azure Application Insights, refer to `https://docs.microsoft.com/en-us/azure/azure-monitor/app/app-insights-overview`.

Azure Application Insights provides a lot of metrics that include, but are not limited to, request rates, response times, failure rates, dependency rates, page views, load performance, and several others.

Implementing distributed tracing

Distributed tracing is also known as *distributed request tracing*. It is a monitoring mechanism that logs traces of the request flow path. Distributed tracing allows you to track how each individual request is handled and helps you locate problems. It is one of the important components when considering monitoring in a microservices architecture. It can be implemented with a centralized logging mechanism where all the traces from different services can be dumped and then monitored. However, with microservices architecture, this is sometimes challenging since the system is based on polyglot technologies, and the monitoring system does not provide the option to integrate with all the platforms or technologies used to build microservices.

Azure Application Insights is one of the cloud services that is best suited in this scenario as well, as it provides integration options with almost all existing popular technologies today. Whether it is .NET, Java, Python, JavaScript, Angular, React, PHP, Node.js, or any other language or platform, it provides an SDK that can be used to log traces.

Some of the practices to consider when implementing distributed tracing are as follows:

- Each request should be assigned a unique number. In a microservices architecture, we can assign this number at the API gateway level.
- Pass the request ID to all the services part of the request context.
- Include the request ID in all the log messages for better traceability.
- Capture information such as start time, end time, and the operation performance, along with other important attributes.
- Log everything centrally to one monitoring platform.

Implementing health checks

Health checks are used to provide a health status for services. A health check is a REST API that is used to check the health of microservices. When developing microservices, you can expose some web methods to return the health status of the service. Health checks can access these endpoints to see if the service is running and validates all areas that include database connections, services endpoint connections, and other resource availability. The web methods should check all this information and return the response. The following figure demonstrates the implementation of a health check API, which calls multiple APIs to check the health status:

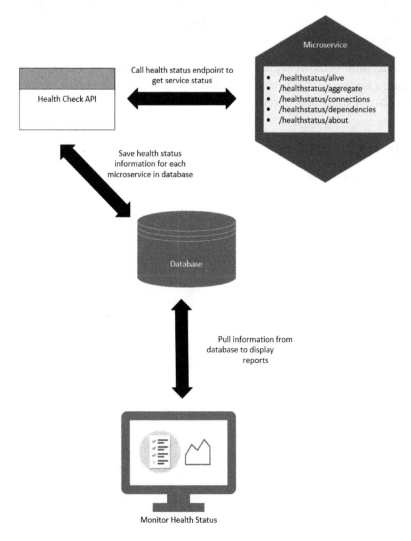

Figure 7.12 – Implementing a health check in microservices architecture

You can also develop a background service that periodically calls the health check API to check the microservices' health status and log into the database. Having said that, you can also build a dashboard to display all this information in a graphical representation.

Summary

In this chapter, we learned about the importance of microservices chassis patterns and their use cases. We discussed some common needs of microservices applications and then moved on to discussing the cross-cutting concerns and pitfalls focused on security, resiliency, idempotency, protocol translations, message transformations, configuration, service discovery, service templates, and monitoring, and followed this by discussing the alternative patterns and practices.

The next chapter will focus on deployment pitfalls, and why DevOps should be adopted before starting a microservices project to automate the build and deployments of services. We will also discuss some best practices, along with examples.

Questions

1. Why is it important to consider the microservices chassis pattern when building microservices?

2. What is the difference between client-side discovery patterns and server-side discovery patterns?

3. Which service in Azure can be used to monitor your live applications?

Section 3:
Testing Pitfalls and Evaluating Microservices Architecture

In this section, you will learn about deployment pitfalls, testing pitfalls, and how to address them by implementing various alternative approaches. You will also learn about evaluating the microservices architecture when building or replatforming existing applications to a microservices architecture.

This section comprises the following chapters:

8

Deployment Pitfalls

Establishing deployments for microservice-based systems requires careful thought and attitude. Not only are microservice-based systems broken into separate services, but they often use various technologies. Unlike monolithic applications, where deployment can be completed in a single step, this paradigm entails implementing a proper strategy. When setting up this strategy, it is essential to learn about the different deployment pitfalls that could hinder the process and how to avoid them throughout the journey.

In this chapter, we will cover the following deployment pitfalls, along with the recommended practices that should be considered when setting up your deployment strategy:

- Failing to establish a deployment strategy
- Using outdated tools and technologies
- Failing to obtain cooperation from all teams
- Choosing your infrastructure before knowing the architecture
- Not considering **Infrastructure as Code (IaC)** for environment setup
- Not knowing the core principles of DevOps
- Deployment stamps

- Deployment rings
- Geode pattern

By the end of the chapter, you will have learned about the deployment pitfalls and how to address them by defining an appropriate deployment strategy and automating deployments using CI/CD.

Failing to establish a deployment strategy

With a deployment strategy, you deploy or update versions of your application in the respective deployment environments. Microservice-based applications provide better agility and speed than traditional applications. Developers can make changes quickly, and with minimal testing, a new version can be released. However, any code change that hasn't been tested properly and has code quality issues will be open to failures and may affect the overall business. To mitigate these issues, developers should define a deployment strategy to check the quality of the application before deployment and must ensure that it is ready to be deployed. Let's look at some of the deployment strategies that you can choose for your organization.

Rolling deployment

The rolling deployment strategy updates all the running instances of the application with new instances. This is one of the easiest deployment strategies and is where you deploy to all the nodes where the application is running incrementally. When you use Kubernetes to orchestrate your containers, Kubernetes is responsible for updating all the containers running within Pods to the most recent version. However, if the new version runs into issues and fails, all the instances will fail because the same version will be used everywhere.

Canary deployment strategy

In the canary deployment strategy, the application becomes accessible to the users in phases. That is, after deploying a new application version, not all traffic is directed to the new version; instead, a portion of traffic is directed to the new version, and then it gradually increases. This method is less prone to risks, and you have more control.

The following diagram depicts the canary deployment strategy, where the microservices will be deployed and released in phases. This helps you test the new versions with real users and evaluate how the new version is working:

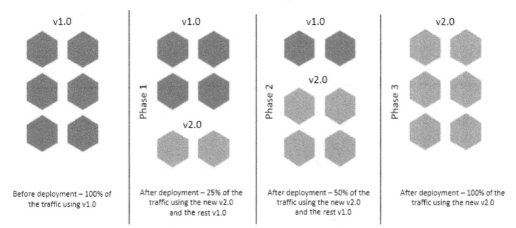

Figure 8.1 – Canary deployment strategy

On Kubernetes, we can implement the canary deployment strategy by configuring the Ingress controller to assign a percentage of traffic by setting some routing rules. These routing rules help increase the traffic to the new service gradually, by directing some percentage of the traffic to new services and the rest to the old services. Also, consider Azure App Service, which enables you to deploy new versions on different deployment slots while monitoring and gradually increasing the percentage of traffic.

> **Note**
>
> Refer to the following link to learn more about Azure App Service deployment slot configuration: `https://docs.microsoft.com/en-us/azure/app-service/deploy-staging-slots`.

In the next section, we'll explore the A/B testing deployment strategy and how it differs from the canary deployment strategy.

A/B testing deployment strategy

A/B testing is similar to the canary deployment strategy, which means that different versions of the same service run simultaneously. However, the new features are enabled or disabled using feature flags. Feature flags can be implemented in your service code to enable a block of code. If a particular feature is enabled, you can control this from the admin interface. If a feature flag is enabled at a certain period, users can access the new features, and then the teams can monitor how their new version is behaving. The following diagram depicts A/B testing:

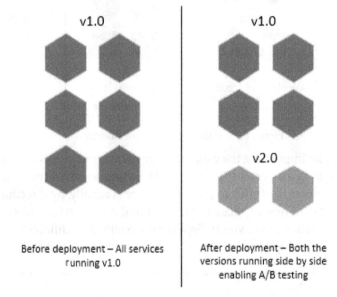

v1.0	v1.0
	v2.0
Before deployment – All services running v1.0	After deployment – Both the versions running side by side enabling A/B testing

Figure 8.2 – A/B testing deployment strategy

Many tools support implementing A/B testing using the feature flags approach. LaunchDarkly is one of the most popular tools that allows you to implement feature flags easily and smoothly.

> **Note**
>
> To learn more about LaunchDarkly, refer to the following link:
> `https://launchdarkly.com/`.

Blue/green deployment strategy

The blue/green deployment strategy necessitates running two production environments concurrently. Here, one is considered a stable environment, known as blue, while the other is considered a staging area for deploying new versions, known as green. You can easily deploy new versions to the green deployment and then roll them out to the blue environment once they are stable. Rolling out a new version can be achieved either by deploying new versions or by switching the traffic to the new environment.

The following diagram depicts a blue/green deployment strategy:

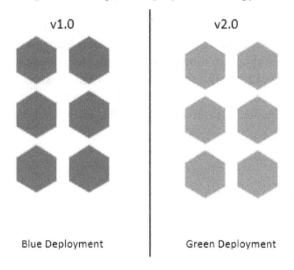

Figure 8.3 – Blue/green deployment strategy

Blue/green deployment is simple to set up and poses less risk as it provides a dedicated environment for testing new versions. Testing is simple and rolling back does not affect the business. Only when the service is stable is it moved to the blue environment.

On the other hand, creating an environment that's similar to blue (production) is complicated, and testing may not cover all the aspects that the system does in a live environment. To implement an environment that's similar to the blue one, you need to double the resources, though this means that additional costs will be incurred.

Traffic shadowing deployment strategy

With the traffic shadowing approach, all the requests that come to the blue environment are routed to the green deployment as well. Unlike the blue/green deployment strategy, a specific percentage of traffic *isn't* routed; instead, all the traffic is mirrored across both environments. This allows us to monitor all the scenarios and how they operate in the new version. There are a few considerations to be made to ensure that if the green environment performs some work, it will not affect the production data. It is recommended to ensure that all the endpoints, database connections, and other configuration values are referring to the green environment and not touching the production environment.

Service meshes help you invoke services that are hosted as part of the green environment. Here, you can use tools such as Istio and Envoy. These tools help you shadow the traffic automatically and provide some additional information, mentioning whether this request is shadowed or not. Envoy adds `-shadow` as the suffix to the hostname. So, for example, if the hostname is `orderservice:5000`, the shadow host becomes `orderservice-shadow:5000`. The following diagram depicts a traffic shadowing deployment strategy:

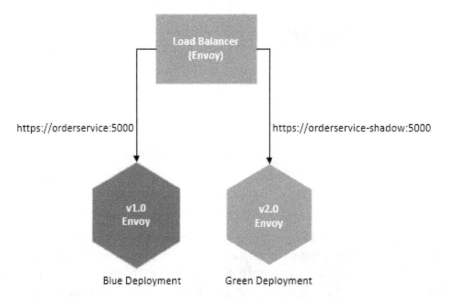

Figure 8.4 – Traffic shadowing

Here, the load balancer is using Envoy to route traffic to both environments at the same time. All the requests that go to the blue environment go to the green environment as well, and the new version can be tested side by side.

Deploying Pods in Kubernetes

Kubernetes is one of the pioneer platforms for orchestrating your containers and works best for production workloads. In Kubernetes, you can deploy and provision various kinds of resources. The deployment resource in Kubernetes holds pod-related information, as well as the replica set, which determines the number of containers that need to be provisioned.

These resources can be provisioned either from the K8s portal or by using the `kubectl` command. The `kubectl` command interacts with the K8s API server, which schedules the task of provisioning that resource. When the deployment is executed, the PPods get provisioned in the K8s cluster, as per the desired state defined in the deployment. When you redeploy the deployment and change the configuration so that it refers to the new image version, K8s performs several checks on the Pods, such as *liveness checks* and *readiness checks*, to name a few. The liveness check determines whether the Pod should be automatically restarted if your application is deadlocked. On the other hand, the readiness check determines whetheryour application is ready to serve.

The following diagram shows how to deploy `order-service` to three Pods inside the K8s cluster:

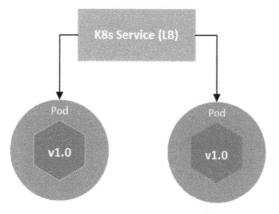

Figure 8.5 – Traffic shadowing

Now, if you want to deploy v2 of `order-service`, you can provision a new deployment object. In this case, it will create the first Pod (the fourth Pod in the cluster) for version 2.0 and then perform the liveness and readiness checks. If both checks are successful, one Pod will be deleted and a new Pod for version 2.0 will be created. This cycle goes on until all the Pods, as per the replica set, get provisioned and the old ones are deleted. A delay may occur between Pod creation since these checks may take some time to ensure there will be no downtime.

The following diagram depicts the rolling deployment pattern in Kubernetes:

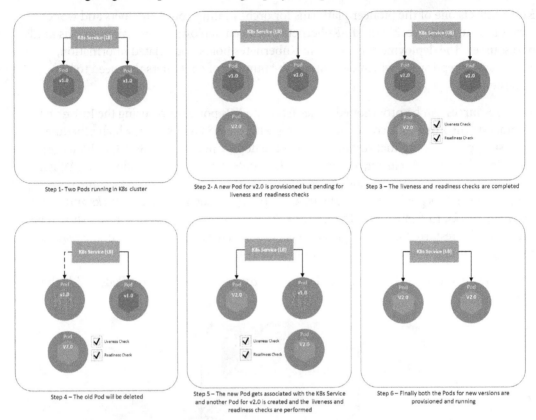

Figure 8.6 – Rolling deployment pattern in Kubernetes

This is the default approach that's used in K8s deployments. However, the canary setup can also be implemented to phase out multiple versions for a different set of users for live testing.

Choosing the appropriate tools and technologies based on the need and breadth of the requirements is essential. In the next section, we'll look at why obsolete tools and technology slow down the microservices journey and how to minimize it.

Using outdated tools and technologies

Technologies degrade over time, and new technologies that are more robust, versatile, and risk-free are introduced. Using outdated tools and technologies increases risks to your business. Always use technology that supports better containerization and deployment options, as recommended for microservice-based architectures. When selecting a specific version, go with the most recent, or N-1. When it comes to deploying microservices, one of the most important requirements is DevOps. Select a tool that includes options for automating development and deployment processes. Some of the tools that can help you execute CI/CD are Azure DevOps and GitHub.

Failing to obtain cooperation from all teams

DevOps embraces collaboration between developers and the operations teams. Traditionally, when applications are deployed, the developers create a build or a package from their machine or a build server, and then they hand it over to the operations team to deploy it on the production server. The operations team follows the deployment manual and deploys the application. You might have seen how an application fails when even a single step is missed. When a part of the deployment (or even sometimes with applications) isn't tested properly, errors or bugs may occur.

In a microservices architecture, we are not talking about one or two applications. We can have more than 50 services that have been independently deployed and are running on different servers. Note that manual deployment is not an option here. DevOps can help establish collaboration between teams and provide visibility to both teams about how the application is built and deployed. Both building and deployment can be automated by creating pipelines that define the list of tasks needed to be executed to build, test, and deploy. Setting up the pipeline's initial structure may take some time but once that is set up, teams won't face issues concerning builds and deployments. Moreover, Azure DevOps provides various other options for controlling deployment by adding pre-approvers, post-approvers, quality gates, and so on.

Infrastructure setup is one of the core principles of the **Cloud-Native Computing Foundation** (**CNCF**) and plays a vital role in the microservices architecture as well. Companies that are coming from the experience of monolith applications or legacy systems may not realize the importance of the right infrastructure for the right implementation. Adopting similar services may not be a good fit for a microservice-based application.

For instance, microservices embrace containerization. This is important if you wish to remove deployment risks and run applications or services isolated in their own environments. Moreover, it also answers the problem of something running fine in development and failing in production. If companies are targeting Microsoft Azure for their cloud platform, they can use Azure Service Fabric or Azure Kubernetes Service to run and orchestrate their containers. However, if moving to the cloud has some restrictions due to company policies, you can set up Service Fabric locally on your on-premises servers or set up Kubernetes and manage it from Azure Arc for Kubernetes.

Not considering Infrastructure as Code (IaC) for environment setup

Setting up the environment is an essential activity to be carried out to deploy any application. Many companies have strategies for creating multiple environments, such as staging, canary, and production to help deploy their workloads for testing and going live. **Infrastructure as Code (IaC)** provides a platform for creating the script for your infrastructure and deploying it to as many environments as necessary. Imagine that you manually set up an infrastructure that takes 5 days to complete for a single environment, while replicating it for other UATs, staging, and performing a canary deployment needs an additional 15 days. Moreover, if any part of the environment crashes, resetting would take the same time again. With IaC, since the whole configuration is written in the form of a script, you can deploy in minutes and easily replicate to multiple environments.

Azure Resource Manager (ARM), from Microsoft, and Terraform, from the open source community, are two of the most popular tools available today that allow teams to build IaC in a declarative manner. Additionally, they allow you to integrate with Azure DevOps to kickstart deployment on demand, when it's scheduled, or on a successful build. Both ARM and Terraform use JSON format to define the configuration, and many tools can be used to run those scripts, such as VS Code, Visual Studio, PowerShell, the Azure console, the Azure CLI, and more.

Some of the properties of these tools are as follows:

- **Template-driven**: All the configuration is defined in a JSON-based template.
- **Declarative**: Uses a declarative model instead of an imperative model.
- **Idempotent**: Resources are not provisioned again when redeploying the template again and again if they have already been provisioned.

- **Multi-service**: A single template can be used to deploy multiple services at once.

- **Extensible**: You can extend the template by adding more resource providers and resources.

For ARM templates, Microsoft has recently built a language known as Bicep. Bicep is a **domain-specific language** (**DSL**) that uses declarative syntax to deploy resources on Azure. The syntax is more concise compared to standard ARM syntax and adheres to type safety, which makes it less complex and simplified.

> **Note**
>
> To learn more about and to get started with Bicep, check out the following link:
>
> ```
> https://docs.microsoft.com/en-us/azure/azure-
> resource-manager/bicep/overview
> ```

Not knowing the core principles of DevOps

DevOps allows the development and operations teams to collaborate, but many companies don't use this ability completely. DevOps helps app innovation by shortening the cycle time and providing results faster. You may utilize DevOps to automate the process between development and operations, allowing you to release software faster to market while also learning, monitoring, and measuring the application in production. DevOps also supports agile approaches that provide value to consumers by allowing teams to plan, execute, and monitor their work continuously. It also allows you to optimize IT resources and increases developer agility.

Azure DevOps facilitates all the stages of the software development life cycle. When the project starts, the backlog can be maintained using various work item segregation tools such as portfolio backlogs and product backlogs. You can create epics, features, user stories, tasks, and bugs to document the task that needs to be performed. It also provides many options including (but not limited to) team capacity, Spring planning, task assignments, and so on. Repos is a popular feature in Azure DevOps that allows users to keep their source code using either Git or **Team Foundation Version Control** (**TFVC**). The build and release process can be automated with pipelines, while test plans can be used for testing and artifacts can be used to retain local feeds such as NPM, NuGet, and Maven packages.

The following diagram depicts the areas where Azure DevOps can be used:

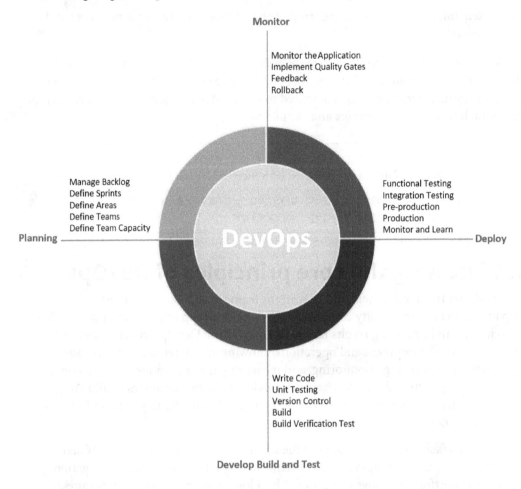

Figure 8.7 – Areas where Azure DevOps can be used

Now, let's look at some of the important features that can be adopted in microservice-based applications specifically.

Feature management

Microservice applications usually follow agile methodologies because they support agile development and quickly releasing to the market. With the agile methodology, the life cycle of releasing a newer version of the application spans over 2 to 3 weeks. In such cases, there could be conditions where the application feature hasn't been completed yet and cannot be shown to the end users. For example, you could have built an e-commerce system but the payment part hasn't been completed yet, where the frontend is working but the backend service isn't.

In this scenario, you don't want to show the frontend option as well and you need to hide this from the web application to avoid users from accessing it. Commenting code and disabling the frontend would require you to uncomment when the feature is completed, which requires deploying the frontend application again. Feature management should be done in such a way that it allows developers to turn the features on and off externally from the application, and then enable and disable them when they want. LaunchDarkly can be used to accomplish this.

Agile practices

As we discussed earlier, microservice applications embrace agility and follow agile methodologies. Knowing about agile practices is an important factor to consider when making teams more productive as it helps them create better quality products and satisfy customers consistently. When all the people involved in the project, such as stakeholders, developers, testers, project managers, and leads, have a shared understanding of the goals, they are more likely to get the desired results.

Agile teams plan continuously throughout the project, which helps them create applications closer to the customer's expectations. To elaborate more about agile, let's discuss some principles that are based on the agile manifesto:

- Customer satisfaction is of utmost priority when building software.

- Embrace change requirements, even late in the development cycle. Agile procedures take advantage of such changes to help customers gain a competitive advantage.

- Deliver functioning software regularly, anywhere from a few weeks to a few months, with a preference for a shorter timeframe.

- Throughout the project, business people and developers must collaborate regularly.

- Build initiatives around people who are passionate about what they're doing. Give them the space and support they require and trust them to do the task at hand.

- Face-to-face communication is the most efficient and effective way of delivering information to and within a development team.

- Working software is the most important indicator of progress.

- Sustainable development is aided by agile procedures. Sponsors, developers, and consumers should all be able to keep up with the pace forever.

- A constant focus on technical excellence and smart design improves agility.

- Simplicity – the technique of minimizing the amount of work that isn't done – is critical.

- Self-organizing teams produce the finest architectures, requirements, and designs.

- The team considers how to become more successful at regular intervals, then adjusts and modifies their behavior in response.

Agile practices embrace microservices development by providing a methodology that minimizes the risk and overhead of large-scale developments by breaking them into smaller work increments and frequent iterations. It supports the prototyping approach by providing a solid collaboration channel for end users.

Rollback strategy

Microservices can be deployed independently, without the need to interfere with other services. It is feasible that one team creating the service will deploy it bi-monthly, while the other team will deliver it bi-weekly. When a new service version is delivered, the service should be self-contained enough that there are no breaking changes.

To understand this concept, let's suppose that we have two services called *Microservice A* and *Microservice B*. *Microservice A* calls *Microservice B* to perform some operation. After 2 weeks, a new version of *Microservice B* is deployed that modifies the signature of the web method *Microservice A* is consuming and breaks the service.

As a solution, versioning should be considered to avoid breaking changes. If a new version of *Microservice B* is deployed, there should be a way *Microservice A* can consume the old version and upgrade it at its own pace. Now, let's suppose *Microservice A* is using the latest version of *Microservice B*, which is v2, and this version contains some issues – in this case, you cannot roll back easily since it breaks the operation *Microservice A* is performing. The rollback strategy should be defined in such a way that the calling service should not fail if the target service version is rolled back. A good practice is to always test your application and set up various deployment environments to test it before releasing it to production.

Using gates and approvals in Azure DevOps

Every deployment undergoes a set of approvals, especially for production apps. Approvals could be pre-deployment or post-deployment and initiate the deployment when it is approved. Azure DevOps provides an easy way to set up pre- and post-deployment approvals and send notifications to the respective participants; once approved, that stage of the deployment is triggered. On the other hand, gates are used to monitor the application. Once the application has been deployed, gates can be used to see how the application is performing and determine whether that release can be promoted to multiple environments or should be stopped.

There are various scenarios where gates can be used:

- To check the work items to see whether any incident or issue has been reported.
- To seek approval from external systems such as Slack or Teams.
- To check the pass rate and other metrics to validate the quality of the release.
- To ensure security scans such as code signing, anti-virus checking, and policy checking.
- To check the telemetry and health status of the application. This can be integrated with Azure Application Insights.
- To wait for change management approval from an external system such as ServiceNow.
- To validate the infrastructure against compliance rules after deployment.

Every business aspires to deploy microservices at an enterprise scale to achieve high availability and scalability. The deployment stamps pattern, which we will explore in the next section, is one of the most important patterns that's used to handle these concerns.

Deployment stamps pattern

The deployment stamps pattern involves deploying multiple copies of the application or service to multiple regions. This is used to achieve high availability and scalability for your applications. If you deploy the application to a single region, you might encounter certain limitations:

- Targeting a single region would impact the business in case of a disaster or if that region goes down for even a few minutes.

- The application is being operated at a global scale, and many customers who are not close to the region where the application has been deployed may face latency issues.

- Segregating the data based on customer location.

- Deploying updates in a controlled manner, such as rolling out region by region.

- Deploying updates based on customer preference and releasing new versions to customers who can handle the update frequency.

As a solution, consider deploying the service to multiple locations, where each can operate and update independently. Each application may scale out within a single region and also offer high availability when deploying it to multiple locations. With stamps, the data is usually sharded, which allows you to keep data independently from other data. Each stamp could be targeting a different version and separate data with a schema or model. You can imagine this as an order service that's been deployed to three different locations, where the orders database is kept separate from each location and may have a different database schema or application version. The following diagram depicts this pattern in detail:

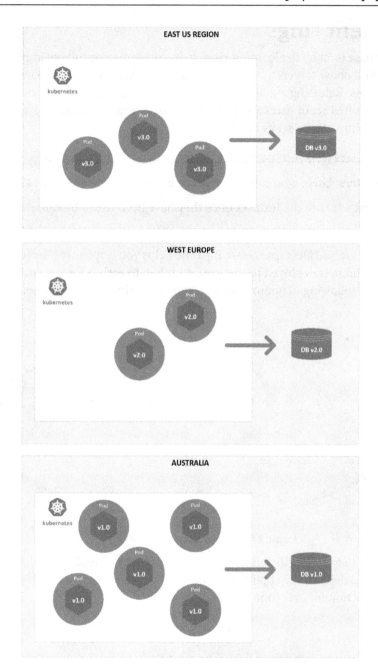

Figure 8.8 – Deployment stamps pattern

The deployment stamps pattern allows you to implement deployment rings, which we will discuss in the next section.

Deployment rings

When the user base is large, deployment rings help you deploy or roll out applications in phases. The first phase targets a small set of users and then increases progressively as the rollout happens. Releasing a major version to a large user base is risky compared to releasing it to a limited set of users and then rolling back when confidence has been built. The deployment rings can be set as follows:

- **Canaries**: Users who voluntarily test the features as soon as they are released

- **Early adopters**: Users who can take risks in using the preview features.

- **Users**: Users who use the features once they have been tested by canaries and early adopters.

Using stages in the Azure DevOps release pipeline helps you implement deployment rings. The release, once built, is deployed to canaries, and then to early adopters, and then to users. The following diagram depicts deployment rings in conjunction with Azure DevOps:

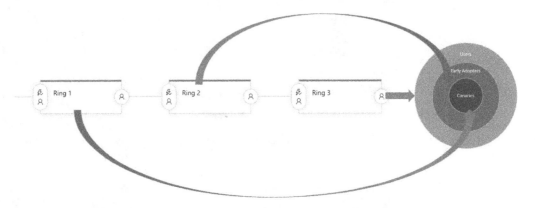

Figure 8.9 – Ring deployments pattern

Here, you can see *Ring 1*, *Ring 2*, and *Ring 3*. *Ring 1* deploys the application for canaries to test and use, *Ring 2* deploys the application for early adopters to use, and *Ring 3* sends the program to production for everyone to use.

Geode pattern

The geode pattern should be used when your applications have users distributed over a wide area and you want to deliver a highly scalable solution. The geode pattern entails distributing a set of backend services over several geographical nodes, each of which can handle any request from any client in any location. When building scalable solutions, data scaling is as important as application scaling. Many companies consider scaling out their frontend apps and sharing a single instance of the data that's been deployed to a central remote location. As a canonical practice, data should be closer to where the application is deployed. The geode pattern improves latency and performance by distributing the traffic based on the shortest path to the closest geode, where each geode is behind the load balancer, such as Azure Front Door or Traffic Manager.

The key difference between the geode pattern and deployment stamps is that with deployment stamps, the data can be fragmented based on the user base. On the other hand, with the geode pattern, the data is duplicated. This design is ideal if a user is a traveler who visits several locations across the world; regardless of where they access the application, they will be linked to the nearest location and have access to the complete data.

> **Note**
>
> To learn more about the geode pattern, refer to the following link: `https://docs.microsoft.com/en-us/azure/architecture/patterns/geodes`.

Summary

We discussed a few deployment pitfalls in this chapter, beginning with the necessity of having a deployment plan, including rolling deployments, canary deployments, and other options, and ending with how Kubernetes deploys Pods. Outdated tools and technologies can also lead to failures; therefore, we should always upgrade and utilize the most up-to-date tools and technologies to avoid making mistakes. Team cooperation is also a significant aspect that influences deployments when it is not well defined. Later we discussed various DevOps practices and some factors to be considered, such as feature management, agile practices, the rollback strategy, approvals, and gates when implementing it. Finally, we learned about a few patterns, such as deployment stamps, deployment rings, and geodes, that we can use to cater to enterprise-scale workloads. With all of these patterns, you can use and adopt the right pattern based on the requirements and implement it to achieve the desired outcome. Testing is critical when developing error-free applications.

In the next chapter, you will learn about testing pitfalls and how to address them by implementing the shift-left testing approach. You will also learn about evaluating the framework that's used when building a microservices architecture.

Questions

1. What is the default deployment strategy that's followed by Kubernetes?

2. What is the main goal of IaC?

3. What are the core practices of DevOps being adopted in a microservice application specifically?

9
Skipping Testing

Why do development teams skip testing? What reason could you have to not test your software? Well, testing is a boring topic for many and something that some folks would rather put off to the last minute. You will hear people say things such as "I just don't have time." Some people blame the business for not wanting to spend the man-hours on testing software, so they never get around to it, or testing gets kicked down the road to the night of deployment. I believe if the business really understood the cost of bugs and the risk of deploying software with bugs, then it would conclude that time must be made for testing software early and often. If you ever want to get the business to listen, then you must place a dollar value on whatever you are trying to get them to understand, be it unit testing, software testing, or adopting agile processes. Businesses are primarily motivated by making and saving money and a solid software testing strategy can help them with both.

A significant question that comes to mind is: Who will educate businesses on the importance of software tests? Will it be the development team or the architect? The answer to the question is the development team and the architect should work together to communicate the importance of testing and educate the business. The architect should include the testing plan in their architectural documentation and during planning discussions with the business.

The business is motivated by money, either saving money or making money, so we must include cost analysis data to prove that, in the long term, costs will be significantly reduced due to a solid testing strategy. We must demonstrate to the business that testing is an investment and not just an expense in terms of time and resources. These savings should be shared with the business leadership to help garner support for management. So, once we have educated the business on the value of testing software, it will see that skipping is not really an option at all.

So, skipping testing is not a good choice and will be disastrous. As a system gets distributed to microservices, testing becomes even more vital to success, as microservices need to integrate with each other and teams need to have a high level of confidence that the added features or functionality will not break their code base or the code base of other dependent services. By testing early, we can avoid disasters, address defects, save the business money, and get to market faster.

Now that we understand the importance of testing and want to avoid the anti-pattern of not testing our software, we will examine a good strategy for testing software, known as **shift-left testing**, and the patterns that support the shift-left approach. We'll be covering the following topics in this chapter:

- Unit testing
- End-to-end testing
- Integration testing
- Load testing

The shift-left testing strategy

Testing earlier and often is the idea behind shift-left testing. By moving testing to earlier in the development life cycle, we will prevent errors from being found later in the process or, worse yet, after deployment into production. Shift-left testing moves testing into the development and build phase, which allows the team to detect defects and bugs early on. This allows the team to address and fix these issues faster and before they can do any real damage.

The shift-left testing approach to software testing greatly enhances the deployment and release process, allowing for faster releases that can occur daily, weekly, or even hourly in some cases.

The lack of testing or not implementing testing properly can lead to defects and bugs being deployed into production, which leads to expensive bug fixes and delays in the releasing of new features.

Let's look at some best practices and strategies in the following sections that will assist the team in implementing a shift-left testing approach, which will lead the deployment team to greater confidence in the code base and allow the business to get to market before the competition with new features. Shift-left testing is win-win all the way around.

Let's understand some important properties of shift-left testing:

- **Planning**: Develop a testing plan that covers the entire development life cycle and make the plan clear for the entire development team to understand their role in supporting the process. Set realistic expectations and goals built to mature incrementally.

- **Testing integrated into the culture**: Testing must become part of the culture enforcement from the top down. Analysis and product owners should follow best practices around creating user stories with good acceptance criteria. Include testers in requirements meetings and in the entire software development life cycle. Embrace test-driven design or behavior-driven design.

- **Standards**: For quality control for code, test user stories. User stories should be complete with good acceptance criteria. Code should be reviewed by testers, senior developers, and peers. Implement good SOLID principles of design. Write code that is testable and easy to change. Last but not least, have standards for tools, frameworks, and third-party libraries.

- **Automation**: This is important for testing and should be implemented early in the development life cycle. There should be automation of testing during the build phase and also during the deployment phase. Use approval gates when necessary when moving from development and testing environments to production. Have employee development pipelines that use the task to automate testing. Testing tools and frameworks should be consistently employed throughout development teams. Run tests using frameworks such as xUnit, NUnit, and JUnit.

- **Agile team structure**: Scrum teams should include testers and teach testers coding skills and start blurring the lines between tester and coders. Sprints should be short, with testers heavily involved in the deployment life cycle.

- **Testers**: Testers should be involved in agile ceremonies such as sprint planning and story refinement. Testers need to learn how to code and, at a minimum, learn to read code; having testers that are comfortable with code is a huge plus. Have developers test and write automation or configure automation with the tester heavily involved in the process.

The testing pyramid

When considering our overall testing strategy, we can create a pyramid to show the level of importance and the amount of effort that should go into each type of testing when it comes to load testing, end-to-end testing, integration testing, and unit testing. Unit testing takes the lion's share of our testing strategy, as demonstrated in the following diagram:

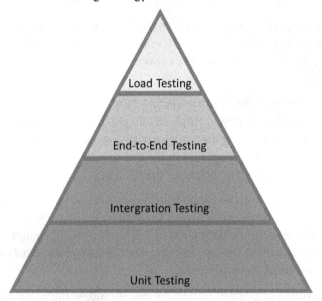

Figure 9.1 – Level of importance of different test types

Unit testing is definitely a shift-left best practice as is starts with the code and the code is about as left as you can get, as it all starts with writing testable, changeable code that meets the acceptance criteria of the business.

Unit testing

Unit testing helps us meet the shift-left approach by moving the testing to the development phase and helps meet the automation portion as well. Developers need to embrace the use of unit testing by using methods such as **test-driven development** (**TTD**), which prescribes writing the test first based on good acceptance criteria. The TTD process has a life cycle of writing a test that fails, then writing the code that makes the test pass, and finally, refactoring the test and the code.

The unit testing life cycle

The unit testing life cycle is made up of three steps; the following figure demonstrates the process:

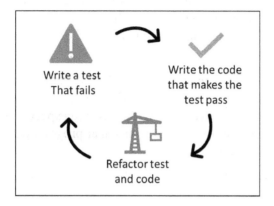

Figure 9.2 – Unit testing life cycle

When deploying distributed systems, having a high level of confidence in your code base is vital and also lets you worry less about introducing bugs that can have a rippling effect across microservices. Unit tests can prove to you that the code is working the way you expect it to and help ensure the correctness of the expected output.

Types of unit testing

There are different types of unit testing. Some are for testing the state of the **system under test** (**SUT**) and some are for testing interactions between dependencies or whether an interaction was invoked:

- **Solitary unit test:** Tests a unit in isolation, mocking dependencies. Testing domain services, business logic, and invariants are solitary tests.

  Testing controllers of RESTful APIs is a solitary test. Inbound and outbound messaging should be tested within the solitary test.

- **Sociable unit test**: Tests a unit and its interactions with dependencies. Testing the persistence layer is sociable test.

Testing collections of value objects is a social test. Testing saga and transaction boundaries is a sociable test.

Scope of units

Business logic in microservices is implemented by using the persistence of entities and value objects, and using sagas to manage those transactions. These services are domain services that are made up of classes. The best practice is to test these classes and their internal methods, which represent their behavior. This is accomplished by creating a test class for each class or SUT and writing a test for each method or function that tests the SUT behavior as expected.

Using a solitary unit test in which an employee mocks the unit's or SUT's dependencies, such as messaging and repositories, will greatly help with test setup and maintenance. A use test case is a set of test input, execution conditions, and expected results developed for a particular objective, such as to exercise a particular program path or to verify compliance with a specific requirement.

Testing should verify the behavior of the SUT or basically the software element or unit being tested.

A unit or SUT can be as small as the testing of methods in a single class or as large as the entire application; it could also involve a collection of classes. It really is up to the team to decide what they define as a unit, but best practices are to keep them as small as possible, focused on a single responsibility. There are some cases where a unit can cross domain boundaries, so careful consideration must be given to deciding the scope of your unit definition.

Strategy

Make unit testing the largest portion of your testing strategy, with the majority of your testing focused on the code and the individual units.

Use TDD, which is important to microservices. By committing to TDD, we will ensure that code is tested as that's the first thing to do after the code is written. It makes sense to write the test first as we need to prove that the code meets the specification of the business and the criteria of the use case. Therefore, having solid acceptance criteria in user stories or requirements documentation allows for the quality control team members to conduct their test and the developers to write code that meets the requirements.

Tools

There are many tools and frameworks to help with unit testing, some for mocking dependencies and some for creating anonymous test data, such as AutoFixture. These tools help you with writing unit tests and with the maintenance of those unit tests as well.

Frameworks and test maintenance

Automation is an important part of the shift-left approach and testing frameworks such as xUnit, NUnit, or JUnit can automate the running of our unit test and can be an important part of our CI/CD pipelines. Testing frameworks can make test maintenance less painful.

The maintenance of unit tests can be a big job and can be costly in terms of time and effort, so I suggest you employ supporting frameworks such as AutoFixture, AutoMock, Mock, and Fluent Assertion.

End-to-end testing

Before we deploy our microservices into production, we need to have a high level of confidence that it will not cause an issue with other services or infrastructure. Although end-to-end testing has value, careful consideration must be given when employing this type of testing as it is difficult to do and costly in terms of time and resources.

We need to be confident that the microservice is behaving as expected. We need to write a test that verifies that the service in question is interacting with infrastructure services and other application and domain services. To successfully test all of these interactions, we need all the services and databases involved to be up and running in an appropriate environment.

End-to-end tests are slow and brittle and can be costly in terms of time and maintenance. End-to-end testing should be minimized and represents the smaller portion of our testing pyramid, as shown in *Figure 9.1*.

The best strategy for end-to-end testing is that it is the smaller portion of our testing. The best practice is to write a journey test that is related to the journey a user takes through the system. For example, we write a test that involves all the operations involved in the user interaction, such as creating an order, updating an order, and canceling an order.

Microservices are often made up of RESTful API services and are owned by different teams. We need to ensure that teams are not breaking other teams' functionality, usually by breaking a contract of an interface, and the services have reliable and stable APIs. We need to make sure that teams do not change their APIs in a way that breaks the API gateway or other dependent services.

A way to verify that all the services can interact is to have all involved services running, and invoke an API or trigger an event that communicates with other services.

When we need to ensure that services are working as expected in microservices, we need to test that each service interacts correctly. We need to do all the following to effectively conduct end-to-end testing:

- All services that the app interacts with need to be running

- Test UI input and correct behavior

- Test UI input and correct output

- The database needs to be running with data seeded

The two types of end-to-end testing

There are two types of end-to-end testing: vertical, and horizontal. Let's learn a bit more about it in this section.

Vertical end-to-end testing is the process of testing the layers of subsystems, for example, testing the backend and data layers, working your way up to the upper layers such as UI.

The advantages to a vertical approach are high test coverage and faster test execution. The test is more focused and provides a layer of safety to critical system functionality.

The focus with vertical testing on the architecture is an area where you can take advantage of **test-driven deployment (TDD)** or **behavior-driven deployment (BDD)**. As with all testing, ensure you include stakeholders such as testers, business experts, developers, and product owners in developing your end-to-end testing to determine what is important to test and how it should be tested. This is important due to the fact that we are focused on business rules, logic, and invariants. Next, we will talk about testing the user journey through the application and how horizontal end-to-end testing helps in facilitating that approach.

Horizontal end-to-end testing, on the other hand, is the process of testing a workflow or transactions that flow through multiple microservices that are involved in the workflow from beginning to end and checks to make sure each microservice correctly behaves as expected. This can include external endpoints or microservices out of our control, but we need to test the output of that external resource as well. External resources add another layer of complexity to end-to-end testing and require special planning and consideration.

There are some advantages to horizontal testing, for example, the test is focused on the user's journey or perspective. Another advantage of this type of testing is being able to discover issues with the user experience or bugs, helping to identify these types of issues prior to release to production. Horizontal testing validates and ensures that business invariants are not broken, and that business logic is covered and tested.

Horizontal tests run across multiple microservices so all those microservices need to be up and running at the time the test executes. This is what makes this type of testing complex and why end-to-end testing should be a smaller portion of your overall testing strategy.

Horizontal testing can include external endpoints or microservices out of our control, which leads us to our next topic on types of environments.

White box and black box environments

In the white box type, testing users can see what is going on, have a UI to interact with and perform actions, and can see whether results have met their expected outcomes.

During white box testing, there are some actions that are not visible to the user, such as calls to backend services that will return the results the user expects. These invisible services are called black box environments.

These backend services are called a black box environment because the actions are said to be performed in the dark, out of the view of the tester or user.

The best practice for horizontal testing is to write a journey test that is related to the journey a user takes through the system instead of writing a test for each operation.

Journey testing

So, for journey testing, write one test that includes the operations involved. For example, if it's an ordering system, test creating an order, canceling an order, and updating an order, which represents the journey a user takes.

An end-to-end test is an acceptance test and should be treated as such. Best practice is to use Gherkin and some type of framework such as Cucumber or SpecFlow to run your acceptance test.

Tools

Docker Compose can provide a way to get all the services up and running.

Tools and frameworks can make end-to-end testing easier to manage. Use SpecFlow, Cucumber, or Cascade to automate these tests.

SpecFlow and Cucumber are behavior-driven design frameworks that work with existing testing frameworks such as xUnit, NUnit, and JUnit. They run a test that is based on the Gherkin language.

Gherkin is a human-friendly language and uses a set of special keywords as follows:

- GIVEN is the context or state the system under test is currently in for this scenario.

- WHEN is the action taken by a user.

- THEN is the expected outcome of the action.

These keywords give structure and meaning to an executable specification. They are steps that form a scenario and are translated into step definition that is expression and combined with a test method. Consider the following example of Cucumber framework step definition:

- Scenario: Products selected

- Given: I have 3 products

Considering the given information, the step definition with an expression and a supporting Java method is as follows:

```
@Given("I have {int} products")
public void i_have_products(int int1) {        }
```

With this, we conclude our discussion of end-to-end testing. Another type of testing that is similar but more focused on a smaller scope than end-to-end testing is integration testing. It is deployed for testing the interaction between two separate microservices.

Integration testing

A microservice needs to interact with other microservices, so testing these interactions is vital but presents some challenges. With integration testing, the primary challenge is that we need to test-run the instances of all the involved services (for instance, when testing a Web API that interacts with a database). This creates the need for complex setup and maintenance of these tests. In the next section, we will explore strategies to address these challenges.

Strategy

Testing interactions between microservices is vital. Integration testing of microservices presents some challenges. We will explore some strategies you can employ. We need to test things such as gateways and the consumption of RESTful API endpoints and domain event messaging being properly consumed and whether the data and the state produced by these services are properly persisted into a database. Unlike end-to-end testing, we do not need all the services to be running; we just focus on a smaller set of interactions, such as the interaction between a repository class and an order class, testing to make sure the test order was persisted properly.

Microservices often need to persist data in a database-like aggregates such as an order, in a document, or in a relational database of some type.

We also have a service that employs patterns such as CQRS that need to read and write data many times, tested by using in-memory objects.

Docker can be used to containerize the database, which can provide the integration database for testing. While you can use this and mocking methods to test integration, they can be difficult to manage and maintain. This brings us to the best practice of using the concept of contracts for testing the integration of microservices. Let's look at contracts and how they can help us test microservice integration.

Contract testing

A **contract** is a concrete example of an interaction between two services. The way we build these contracts depends on the type of interaction involved. Contract testing has two types of contracts: consumer and provider, or producer.

The following figure illustrates the contract testing steps and the interactions between the consumer and the provider:

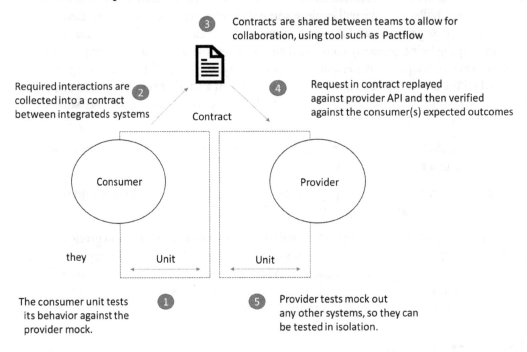

Figure 9.3 – Contract testing

Contacts are for both consumers and providers, which ensures that they are in agreement on the API level. Each contract's interactions with consumers or providers are tested in different ways:

- Consumer tests use the contracts to configure stubs (returns value) that simulate the provider, enabling you to write integration tests for a consumer that does not require a running provider.

- Provider tests use contracts to test the SUT, using mocks for SUT dependencies.

- Contract testing provides a mechanism for mocking and stubbing or integrations. We can use Pact.io to provide the capability of testing integrations between services and repositories. Contract tests can provide a test for HTTP requests or messaging approaches and are called pacts:

  i.   **HTTP pact**: An expected request is what the consumer is expected to request. This in turn has a minimum expected response, which is the parts of the response the consumer wants the provider to return.

ii. **Messaging pact**: Contracts support modern distributed systems that use messaging frameworks such as RabbitMQ, Kafka, or Azure Service Bus. These types of interactions are called "message pacts." This is achieved by abstracting the frameworks out and just focusing on the message being passed between.

Tools

Contract testing is the best practice for microservices integration testing and you can use frameworks such as **Pact.io** or **pactflow.io** to create a contract file that is a concrete JSON document that provides the specification of the provider and the consumer.

Pact is good for designing and testing integrations where teams in your organization control the development of the consumers and the providers, and also where the consumer's team will drive the provider's team requirements or features. So, in short, Pact is best for white box integrations where the teams have a good understanding of the system under test.

I encourage teams to explore contract testing for integration. To really explore contract testing, go to Pact.io for more information on Pact.io and Pactflow.io.

We have explored integration testing and some best practices around it. Now we need to ensure that our application can meet the demands of our users and provide a good user experience.

Load testing

Load testing can help identify issues with horizontal and vertical scaling bottlenecks and issues with threading or connection pools, to name a few. It can inform us on how we need to scale our applications or help identify any blocking calls for async operations. Load testing can verify latency as requests and responses flow through the system network.

Microservices may have a dependency on other services or resources and these services can have varying degrees of performance. Some may be fast and sometimes some services can have an issue and run slower than expected. We need to know how the system will react to these varying degrees of performance and how it will impact the service we are testing.

The microservice environment is dynamic and constantly changing. We need to test our services to see whether they can handle this dynamic nature of microservices, and how all of this will handle the load of active users in a production environment.

By load testing, we can identify weakness or bottlenecks in our system and expose our services' ability to scale and avoid untimely crashes due to some issues with the code base that were not identified during development. Unit tests and other testing can miss an issue with threading or connection pools, for example, so load testing can help identify these issues and correct them before we deploy. We need to test the anticipated demand. We need to understand what kind of demand we expect and our test should address the need to meet such a demand on our system. Let's explore some strategies and best practices around load testing.

Strategy

A strategy is important to leverage tooling for load testing to help with monitoring and metrics and with virtualization and test execution and feedback from the test:

- **Risk**: Focus on high-risk parts of the system first and then work on other parts of the system as required.

- **Virtualize**: Mocking services using virtualization use containers and services that provide virtualized endpoint URIs. Using tools and services to provide this virtualization can help the team conduct and observe these types of tests.

- **Metrics and monitoring**: Use tools such as JMeter, Grafana, and Gatling to monitor and provide metrics and execution mechanisms so you have visibility of the performance of requests and responses.

Tools

There are many tools out there for load and performance testing but here are a few that may be helpful to explore. It is highly recommended to use tools such as these in your load testing strategy:

- **Apache JMeter**: Open source software designed for load testing and performance testing

- **Grafana**: View time series data and telemetry data; gives a view of data collected from Prometheus

- **InFlux**: Time series, operations monitoring application metrics, Internet of Things data

- **WebLOAD**: A performance and load testing solution; provides test planning, test design, test execution, and performance analysis

- **Gatling**: Designed for continuous load testing and integrates with deployment pipelines

Testing microservices is vital to successfully develop and deploy microservices. Testing should start early in the development life cycle and should start with the code and continue through each phase from the build to the release phase.

Summary

So, to wrap things up, we saw how important testing is to develop and deploy microservices and how this can help avoid deploying microservices that have bugs that could cause costly issues in production or, even worse, complete failure. Skipping testing is not a good option and can lead to unnecessary failure of your microservices so shift your testing to the left, start testing early, and often through all phases of the software development life cycle. We have explored some of the shift-left testing strategy and the types of testing that support this approach.

Unit testing the code is where it all starts and should represent the majority of your testing. We explored end-to-end testing and its complexity and how we should use the user journey through the app to guide this approach. Integration testing helps test interactions between services and ensures the correct behavior is evoked. Similarly, we learned about performance testing and load testing and understood their importance to make sure the application can handle increased demand and provide a good user experience. Testing is a must for successful microservice implementation and should be shifted to the left in the development lifecycle. Skipping testing will only lead to failures, in my opinion.

In the next chapter, we will discuss the factors to consider when developing microservices to serve as a checklist to help you on your journey with microservices.

Questions

1. What is the key to effective unit testing?
2. What important considerations must be made before conducting end-to-end testing?
3. What is the key to testing integrations between microservices?
4. What is the best load testing strategy?
5. Why is testing important to microservices in particular?

Further reading

- **Load Testing .NET Core**: `https://docs.microsoft.com/en-us/aspnet/core/test/load-tests?view=aspnetcore-5.0`

- **Integration Testing**: `https://docs.microsoft.com/en-us/aspnet/core/test/integration-tests?view=aspnetcore-5.0`

- **Performance testing**: `https://docs.microsoft.com/en-us/azure/architecture/framework/scalability/performance-test`

- **Testing tools**: `https://docs.microsoft.com/en-us/azure/architecture/framework/scalability/test-tools`

- **Testing Checklist**: `https://docs.microsoft.com/en-us/azure/architecture/framework/scalability/test-checklist`

- **JMeter**: `https://docs.microsoft.com/en-us/azure/architecture/example-scenario/banking/jmeter-load-testing-pipeline-implementation-reference`

10
Evaluating Microservices Architecture

Many organizations today are building or replatforming applications into a microservices architecture. When decomposing any application into a microservices architecture, we need to understand which factors are important when making decisions and to analyze if that application is the right candidate for a microservices architecture.

In this chapter, we will summarize the important factors that will help you to build knowledge and understanding of the assessment of a microservices architecture. These factors are equally important when you are building a new application or transforming an existing application into a microservices architecture. You can start your discussion with the customer and refer to these factors to drive your discussion and make a decision.

Here are the factors we will discuss in this chapter:

- Identifying the core priorities of a business
- Managing architectural decisions
- Team structure
- Choosing the right methodology

- Decomposition strategy

- Evaluating the **development-operations (DevOps)** capability

- Understanding which part of the business is going to change more rapidly

- Infrastructure readiness

- Release cycles

- Communication protocol across services

- Exposing services to clients

- Distributed transaction handling

- Service deployment

- Capabilities of the hosting platform

- Deployment strategy

- Monitoring services

- Assigning a correlation token to each service request

- Defining a microservices chassis framework

- Shift-left approach for testing and security

In this chapter, we will discuss some key aspects to think about when analyzing a microservices architecture. Let's get started!

Identifying the core priorities of a business

To start the evaluation of a microservices architecture, you need to understand the priorities of the business to make sure that the architecture is fit for its purpose. The areas you should discuss with the organization while assessing the microservices architecture are listed here:

- Reliability

- Innovation

- Efficiency

To understand the priority of these areas, you can ask the team about how much importance they want to give to this area on a scale of 0 to 10. Moreover, you should also document the expected **service-level agreements (SLAs)** associated with different components of the application to ensure that it meets organization commitments. These priorities and commitments serve as an anchor to guide our assessment.

Managing architectural decisions

A microservices architecture thrives when you allow teams to have autonomy. This doesn't mean that teams can do anything, but it gives them the freedom to make decisions under an umbrella of formally agreed principles. This is called shared governance, which states the consensus across the teams on how they want to address the broader architecture strategy for microservices.

The following aspects are important when evaluating this factor:

- Do you have shared governance in place?
- Are you maintaining an architecture journal along with your microservice to track decisions?
- Are your architecture journal documents all your architectural decisions?
- Is your architecture journal easily accessible by your team?
- Do you have a framework in place for evaluating tools, technologies, and frameworks?

Team structure

Having the right team structure is very important for microservices to avoid unnecessary communication across teams. A microservices architecture promotes the formation of small, focused, and cross-functional teams to deliver business capabilities. This requires a shift in mindset, led by team restructuring.

The following aspects are important when evaluating this factor:

- Are the teams segregated based on subdomains, where one subdomain is owned by only one team?
- Are these teams cross-functional in that they have sufficient capacity to build and operate a microservice?
- How often do your team members spend time on ad hoc tasks related to development or operations?

- Do you have a practice of measuring technical debt?
- How much time do you spend minimizing technical debt?

Choosing the right methodology

Ensure that you have used agile practices for your microservices application, as this comes with lots of benefits. The basic idea behind choosing a microservices architecture, along with many other benefits, is embracing change.

The **Twelve-Factor app methodology** provides the guidelines for building scalable, maintainable, and portable applications by adopting key characteristics of immutability, ephemerality, declarative configuration, and automation. Incorporating these characteristics and avoiding common anti-patterns will help us build loosely coupled and self-contained microservices.

Implementing these guidelines will help us build cloud-native applications that are independently deployable and scalable. In most cases, failed attempts at microservices are not due to complex design or code flaws but to incorrectly setting the fundamentals from the start by ignoring widely accepted methodologies.

Decomposition strategy

Choose something that's not too critical to the organization; once you are ready, move on to something that's going to bring the most value. Transforming or replatforming a monolithic application to a microservices architecture is not a straightforward process. Start with the least dependent module or component and extract it to create a new microservice. Sometimes, it may be possible that one module can be further extracted into multiple services. This can be analyzed properly if you know the modeling techniques and can map the business domains or subdomains to microservices. Once the service is extracted, you may need to build an anti-corruption layer, which is a wrapper service to communicate back to the monolithic application to keep other modules or applications in a running state.

For example, in the student management system, we can have modules as student management, course management, attendance management, document management, and employee management. We can start replatforming the application from attendance management since it seems to have fewer dependencies on other modules, whereby students can come and mark their attendance. From the frontend, when the student marks the attendance, instead of calling the monolithic application's module, a code will be written to call the new attendance management service. A second pick could be a document management service that students can use to upload their documents. For this, we may need to write an anti-corruption layer if there is a scenario to update the student's main record when a document is uploaded. You can learn more about the decomposition strategies in *Chapter 4, Keeping the Replatforming of Brownfield Applications Trivial.*

The data decomposition and migration could be challenging with respect to data synchronization, multi-writes to both monolithic and microservice interfaces, data ownership, schema decomposition, joins, the volume of data, and data integrity. Various approaches can be adopted, such as keeping a shared database between microservices, decoupling the databases to a set of services based on business capability or domain, or segregating the databases so that they are completely isolated to the service itself. Keeping the database per service is the most preferred option. However, it is not easily possible in many cases, and in such scenarios, you can implement patterns such as a Database View pattern and a Database Wrapping Service pattern.

Moreover, there are various database technologies in existence today that help you to choose the right technology fitting the requirement. For example, in a shopping cart service, DocumentDB might be the right database technology for keeping cart information because you don't need a relational database and you can keep the cart information as a document indexed by a user **identifier** (**ID**).

Evaluating the DevOps capability

Assessing the DevOps capability is very important when evaluating a microservices architecture. Microservices are meant to bring rapid development and change adoption in an organization and DevOps is one of the core practices that should be in place to achieve this. The following points should be considered when evaluating DevOps capability for a microservices architecture:

- Does the organization understand the fundamentals of DevOps?
- Are DevOps practices implemented?

  - Are agile practices implemented?

- Is **continuous integration** (**CI**) implemented?

- Is continuous delivery implemented?

- Is continuous deployment implemented?

- Is **Continuous Monitoring** (**CM**) implemented?

- Assess if the right tools are used that support implementing CI/CD.

- Assess how the configuration of staging and production environments are kept/ utilized in the application.

- Is an **Infrastructure as Code** (**IaC**) practice being put in place?

Understanding which part of the business is going to change more rapidly

A microservices architecture embraces change. When evaluating the architecture, discuss with the organization which areas they think will change the most often. Building microservices for such areas will allow us to be more agile in responding to client requests while avoiding disruptions to the main application. Unlike a monolithic application, where you make a change at one place, a microservices architecture goes through several stages of regression testing.

With microservices, each service can be tested independently without affecting the main application and requires less effort and time when modification is needed. As you continue your microservices journey, it is important to watch out for any dependency between microservices to help you remodel individual microservices. It is also important to address the changing needs of the business, which may require decomposing a microservice into different microservices following **domain-driven design** (**DDD**) modeling techniques, which are discussed in *Chapter 2, Failing to Understand the Role of DDD*.

Some of the factors you should focus on are listed as follows:

- Is the service independently deployable?

- Does the service follow DDD?

- Does the service follow SOLID principles?

- Is the database private to the service?

- Is the service built using the supported microservices chassis pattern?

- What is the percentage of microservices that map to business domains?

Infrastructure readiness

Infrastructure readiness is a very important factor to consider when moving toward a microservices architecture. If infrastructure is not properly built or if the right tools are not used, it will impact the performance, availability, and scalability of the application. There could be scenarios where an application is built with all recommended techniques and practices but without having a correct infrastructure, giving a bad performance.

The following factors must be considered when evaluating infrastructure readiness in a microservices architecture:

- Does the deployment infrastructure ensure the scalability of the services?

- Does the deployment infrastructure support provisioning through scripts that can be automated using DevOps CI/CD?

- Does the deployment infrastructure offer a proper SLA for availability?

- Is there a **disaster recovery (DR)** plan and drill schedules for the whole infrastructure in place?

- Is the data replicated synchronously/asynchronously to different regions to ensure data availability for the DR environment?

- Is there a proper data backup plan in place?

- Are deployment options documented?

- Does the deployment infrastructure monitor and ensure the desired number of services are running?

- Does the deployment infrastructure support self-healing of services?

- Do you have **Open Systems Interconnection (OSI) Layer 7 (L7)** load-balancing capabilities?

- Do you have a **web application firewall (WAF)** layer in place?

- Do you have throttling policies implemented?

Release cycles

A microservices architecture embraces rapid development and helps to shorten the release cycles for applications or services. The following aspects are important to discuss when evaluating this factor:

- What is the frequency of building or releasing applications?
- Are you following any practice of releasing the application versions—that is, monthly, quarterly, half-yearly, or yearly?
- How frequently does a deployment cause an outage?
- How long does it take to recover from an outage?
- Are the new service versions following semantic versioning?
- Are the versions released first to staging and then production?
- Are the **application programming interfaces** (**APIs**) following proper version guidelines?
- What makes an API version change?
- Have you implemented versioning?
- What is your approach to handling **HyperText Transfer Protocol** (**HTTP**) API versioning?

  - **Uniform Resource Identifier** (**URI**) path versioning
  - Query parameter versioning
  - Content-type versioning
  - Custom header versioning

- Are you performing **Google remote procedure call** (**gRPC**) versioning?
- Have you implemented event versioning?

Communication protocol across services

Microservices are independent fine-grained services that communicate with each other across process boundaries, hosted across the network, to address different business use cases. Choosing the right communication protocol is a very important factor to ensure reliable communication. The following aspects should be considered when assessing this factor:

- Are you following an API-first approach?

- Do you have deep chaining of services over synchronous communication protocols?

- Do you have asynchronous communication anywhere in the system?

- Which message broker technology are you using?

- What is the throughput of messages received or processed by the services?

- Do you have direct client-to-microservice communication?

- Do you need to persist messages?

- Are you using a materialized view pattern to address the chatty behavior of microservices?

- Have you implemented a retry pattern, a circuit breaker, exponential back-off, and jittering?

- Have you defined domain events to facilitate communication between microservices?

Exposing services to clients

An API gateway is one of the core components when developing a microservices architecture. Exposing microservices directly to clients is not a good practice due to many factors. You should evaluate the following:

- Are the services directly consumed by the frontend application?

- Is there a gateway that acts as a facade for all backend services?

- Does the gateway provide quota limits, mocking responses and filtering **Internet Protocol (IP)**, and other policies?

- Are you using multiple API gateways to address the needs of different clients?

- Does the gateway provide a portal where clients can come to discover and subscribe to services such as a developer portal in Azure API Management?

- Does your API gateway provide L7 load balancer or WAF capabilities?

Distributed transaction handling

Distributed transaction facilitates the execution of multiple operations as a single unit of work. In a distributed transaction, a command is an operation that performs a state change and causes an event to occur. These events might alert the next phase or the entire system of the transaction's success or failure. If they are successful, they can cause another command to be performed, which can then cause an event, and this will continue until all transactions are completed or rolled back, depending on whether the transaction is successful or not. Let's explore the important aspects to consider while evaluating this factor, as follows:

- How many distributed transactions exist in the system?

- What is your approach to handling distributed transactions—is it using an orchestrator or choreography pattern?

- How many transactions span over multiple services?

- Are you using a long-chain operation when a transaction spans multiple services?

Service development

With a microservices architecture, the application is decomposed into a set of various services. You can use any technology to build any service. However, each technology has its own challenges and operational characteristics. Because of the spread of these technologies, organizations may miss out on developing internal expertise, and when engineers transfer teams, ramp-up time increases. Additionally, because companies use a variety of languages, creating and maintaining shared library components has become extremely challenging. In practice, businesses tend to focus on a smaller number of technologies.

However, when you are building a similar service using the same technology, you can sometimes reuse the code or framework to reduce your development efforts. For that reason, in some cases, reusing the existing code seems to be a good practice. Here are some important aspects to consider when evaluating this factor:

- Are you maintaining a service template or utilizing a framework to kick start new service development?
- Do you have a single code base for microservices?
- Are you isolating dependencies?
- Are you externalizing configuration?
- Is sensitive configuration securely stored?
- Are you containerizing your services?
- Have you used a cache where appropriate?
- Do you know your data consistency requirements?

Capabilities of the hosting platform

One of the main benefits of building a microservices architecture is to embrace the easy scalability of services. Containerizing a monolithic application is a challenging task. However, if the application is decomposed into fine-grained services, these can be easily containerized and scaled out based on the need. Discussing the capabilities of the hosting platform for a microservices architecture is an important factor. Here are some aspects that you can discuss to evaluate this factor:

- Which hosting platform are you using to deploy your services?
- Is the hosting platform scalable?
- Does the hosting platform support auto-scaling?
- How much time is needed to scale the hosting platform?
- Is there any proper SLA provided by the vendor for the hosting platform?
- Is there any DR site for the hosting platform?

Deployment strategy

A deployment strategy is a method for distributing or upgrading versions of your program on various deployment environments. Traditional applications lack the agility and speed of microservices-based solutions. Developers may make changes fast, and a new version can be issued with minimal testing. The following aspects should be considered when evaluating a deployment strategy:

- Have you documented a deployment strategy for deploying your services?
- Are you using outdated tools and technologies for deploying or hosting your services?
- Is there any collaboration needed with other teams when you deploy services?
- Are you setting up the infrastructure using IaC?
- Have you leveraged DevOps capabilities to automate deployments?
- Do you have a practice of propagating the same build to multiple environments?

Monitoring services

Monitoring is an important factor for any microservices-based application. Since the application is split into a set of various services running independently, troubleshooting errors becomes a daunting task. If you are following proper semantic logging for monitoring your services, your teams can easily monitor and investigate errors and respond quickly. Here are some aspects to be considered when evaluating this factor:

- Do you have a practice of monitoring applications once they are deployed?
- Is there a proper logging mechanism in place?
- Is there distributed tracing infrastructure in place?
- Is there a practice of recording exceptions?
- Are defined health checks implemented for each service?
- Are you following semantic logging?
- Have you implemented key metrics, thresholds, and indicators?

Assigning a correlation token to each service request

A microservices architecture is distributed in nature and many services interact with each other to complete a business use case. A correlation token is a unique string (preferably, a **globally unique ID**, or **GUID**) that is assigned to each request for troubleshooting purposes.

For example, if there is a long-chain operation where many services are involved, passing the correlation token to the service helps to investigate the issue easily if any of the services fail during that transaction. Usually, each service has its own database and keeps the correlation token within the database record as well.

These two things are worth mentioning:

- Have a correlation ID that correlates the end-user request to requests made to various microservices so that we can quickly locate logs in each of the services that were involved in an end-user request.

- The cost to store and make logs searchable can be large and, for that, some companies end up sampling or storing the logs for a shorter period—both approaches have their pros and cons. If sampling is used, care should be taken to ensure that all microservices involved in an end-user request keep logs of the same end-user request or we won't be able to track a failed request.

The following points should be considered when discussing correlation tokens:

- When is the correlation token assigned?
- Who is responsible for assigning the correlation token?
- Do you save correlation tokens in the service's database?
- What is the format of the correlation token?

Defining a microservices chassis framework

A microservices chassis framework is a foundational framework for fundamental cross-cutting issues such as logging, security, exception handling, tracing and distribution, and so on. The following aspects are important to consider when evaluating this factor:

- Do you have a microservices chassis framework in place?
- Are your services secured and do you handle security constraints inside each service?

- Are you using any resiliency patterns for microservices?
- Have you implemented idempotency for your services?
- Are you using any protocol translation as part of your microservice implementation?
- Are you doing any message transformation inside a microservice?
- Have you implemented client-side or server-side discovery patterns?

Shift-left approach for testing and security

A shift-left approach should be adopted early in the development life cycle of microservices applications. In a shift-left approach, you bring different activities (security, testing) into the development cycle earlier on to improve the quality of the application.

For example, if you are building a new application, you should test the security of the application in the design phase, by using tools such as Microsoft Threat Modeling. Moving forward, during development, the code should be security scanned using **static application security testing** (**SAST**), and furthermore, when the application is deployed, it should be tested using **dynamic application security testing** (**DAST**) tools.

The following aspects are important to consider when evaluating this factor:

- Do you follow a practice of implementing unit tests?
- Do you follow a practice of evaluating your architecture at the design phase?
- Do you have any tools to measure code quality?
- Do you have automated tools to test your applications?
- Have you implemented **development, security, and operations** (**DevSecOps**)?
- Are you using any testing frameworks?
- Have you used WAF at the frontend?
- Are you embracing shift-left testing?
- Do you follow a practice of writing unit tests?
- Do you have a practice of doing integration testing?
- Do you have a practice of doing **end-to-end** (**E2E**) application testing?
- Do you have a practice of doing load testing?
- Do you follow the practice of doing automated testing?

- Do you follow the practice of doing penetration testing?

- Do you follow the practice of doing chaos engineering?

- Are tests automated and carried out periodically or on-demand?

- Do critical test environments have 1:1 parity with the production environment?

Summary

In this chapter, we learned about various factors that are important for assessing a microservices architecture. Giving these factors importance while you are building or replatforming an application helps you highlight areas where improvement is needed. Moreover, you can discuss those improvements with teams to increase the maturity state of their architecture.

In this book, we learned about various anti-patterns and architectural pitfalls to avoid when creating a microservices architecture. We hope that you now have a better grasp of these pitfalls and will consider them when developing new applications or replatforming existing applications to a microservices architecture.

Assessments

This chapter has the answers to all the questions at the end of the chapters of this book.

Chapter 1: Setting Up Your Mindset for Microservices Endeavor

1. The design principles of microservices are single responsibility, domain-driven design, encapsulation, interface segregation, the culture of autonomy, ownership, and governance, independently deployable, the culture of automation, designing for failures, and observability.

2. The microservices architecture components are microservices, messages, persistence, state management, orchestration, service discovery, and API gateway.

3. Following the design principles of microservices and adopting the Twelve-Factor App methodology will help to build highly scalable, maintainable, and portable microservices for cloud-native.

Chapter 2: Failing to Understand the Role of DDD

1. Yes

2. Knowledge crunching domain model, use Ubiquitous Language, and bind the model to the implementation

3. Transactional boundary

Chapter 3: Microservices Architecture Pitfalls

1. The Aggregator microservices pattern can help you build microservices that are responsible for orchestrating communication between different microservices. It acts as a single point of contact that's responsible for calling different microservices and gathering data to perform a business task.

2. Teams should evaluate different tools, technologies, and frameworks under a shared governance model with the help of a capability/feature matrix addressing different aspects.

3. Dapr is a portable and event-driven runtime that facilitates different aspects of building microservices-based applications on the cloud and edge. It includes service-to-service invocation, state management, resource binding, publish/subscribe, observability, secrets, and actors as building blocks.

Chapter 4: Keeping the Re-Platforming of Brownfield Applications Trivial

1. The main objective of migrating from monolith to microservices is to enhance scalability and productivity. The application is decomposed into a set of independent services that are easier to maintain and change, embracing agility.

2. The following are the factors to consider when re-platforming a monolithic application to a microservices architecture:

   - Knowledge of business domains

   - Awareness of the change in infrastructure

   - Choosing the right technology and embracing the learning curve

   - Moving to the cloud

   - Understanding the difference between core microservices and API microservices

   - Avoiding chatty services

   - Ensuring development, delivery, and operation readiness

3. They are important factors since all of these can bring value to the business. If the application is available and scalable in nature, it can handle all kinds of workloads, ensuring availability to the users. Reliability, on the other hand, helps to deliver a quality product.

Chapter 5: Data Design Pitfalls

1. True

2. False

3. Single database, two databases read and write, and event sourcing

Chapter 6: Communication Pitfalls and Prevention

1. In synchronous communication, the client sends a request and waits for a response from the service. Synchronous communication can be blocking or non-blocking, but the client can only continue further processing once the response is received from the service. In asynchronous communication, the client sends a message to the message broker that acts as middleware between the client and service. Once a message is published to the message broker, the client will continue processing without waiting for the response. The service can subscribe for specific messages to the message broker and place responses back in the message broker after processing. Both the client and service are unaware of each other and can continue to evolve without any dependency.

2. Direct client-to-microservice communication introduces tight coupling between the client and microservice, where clients need to be aware of how different microservices can work together to achieve a business use case, which results in implementing service orchestration by the client application. Furthermore, calling each microservice from the client adds latency and affects the user experience and mobility.

3. When multiple microservices call each other repeatedly in a long chain, there is potential that a microservice may take more time to respond than expected, which can result in timeouts. These timeouts will initiate additional retry requests with an expectation that the operation may succeed and the flood of these retries will eventually make the system unusable. The situation is known as a retry storm.

4. There are four different approaches to HTTP/REST API versioning, as mentioned next:

    a) URI path versioning

    b) Query parameter versioning

    c) Content-type versioning

    d) Custom header versioning

5. A service mesh provides an abstraction layer on top of microservices to provide capabilities such as traffic management, security, policy, identity, and observability for your microservices. The idea behind the service mesh is to decouple these concerns from the implementation of microservices and handle them in a consistent manner across the microservices architecture.

Chapter 7: Cross-Cutting Concerns

1. It provides a set of building blocks that helps you to use them instead of building it from scratch and helps you quickly and easily get started with microservices.

2. With the client-side discovery pattern, the client is aware of the service registry and queries the service registry to know the endpoints of services. Whereas with the server-side discovery pattern, the client makes a request to the load balancer, which routes the request to the service registry to know the endpoints of services to consume.

3. Azure Application Insights.

Chapter 8: Deployment Pitfalls

1. The rolling deployment is the standard default deployment strategy followed by Kubernetes.

2. The main benefit of IaC that it provides a code-based approach that makes it easy to provision/deploy things faster and eliminating the manual process.

3. Core practices are feature management, agile practices, rollback strategy, gates and approvals setup

Chapter 9: Skipping Testing

1. Write the test first using test-driven design and automate the test execution.

2. Minimize end-to-end tests using techniques such as journeys, which test the user journey through the system.

3. Use contract testing to mock and stub out integration tests.

4. Focus on the critical parts of the system first.

5. Testing using the shift-left approach by testing early and employing unit testing and integration tests and load testing will give a high level of confidence that our microservices will provide a pleasant user experience for our users and meet the needs of the business.

Packt.com

Subscribe to our online digital library for full access to over 7,000 books and videos, as well as industry leading tools to help you plan your personal development and advance your career. For more information, please visit our website.

Why subscribe?

- Spend less time learning and more time coding with practical eBooks and Videos from over 4,000 industry professionals

- Improve your learning with Skill Plans built especially for you

- Get a free eBook or video every month

- Fully searchable for easy access to vital information

- Copy and paste, print, and bookmark content

Did you know that Packt offers eBook versions of every book published, with PDF and ePub files available? You can upgrade to the eBook version at packt.com and as a print book customer, you are entitled to a discount on the eBook copy. Get in touch with us at customercare@packtpub.com for more details.

At www.packt.com, you can also read a collection of free technical articles, sign up for a range of free newsletters, and receive exclusive discounts and offers on Packt books and eBooks.

Other Books You May Enjoy

If you enjoyed this book, you may be interested in these other books by Packt:

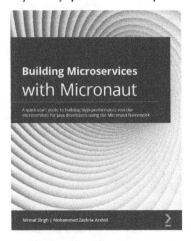

Building Microservices with Micronaut

Nirmal Singh, Zack Dawood

ISBN: 9781800564237

- Understand why Micronaut is best suited for building microservices
- Build web endpoints and services in the Micronaut framework
- Safeguard microservices using Session, JWT, and OAuth in Micronaut
- Get to grips with event-driven architecture in Micronaut
- Discover how to automate testing at various levels using built-in tools and testing frameworks
- Deploy your microservices to containers and cloud platforms
- Become well-versed with distributed logging, tracing, and monitoring in Micronaut
- Get hands-on with the IoT using Alexa and Micronaut

Python Microservices Development - Second Edition

Simon Fraser, Tarek Ziadé

ISBN: 9781801076302

- Explore what microservices are and how to design them
- Configure and package your code according to modern best practices
- Identify a component of a larger service that can be turned into a microservice
- Handle more incoming requests, more effectively
- Protect your application with a proxy or firewall
- Use Kubernetes and containers to deploy a microservice
- Make changes to an API provided by a microservice safely and keep things working
- Identify the factors to look for to get started with an unfamiliar cloud provider

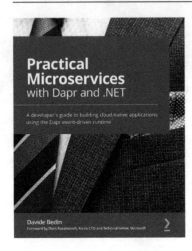

Practical Microservices with Dapr and .NET

Davide Bedin

ISBN: 9781800568372

- Use Dapr to create services, invoking them directly and via pub/sub

- Discover best practices for working with microservice architectures

- Leverage the actor model to orchestrate data and behavior

- Use Azure Kubernetes Service to deploy a sample application

- Monitor Dapr applications using Zipkin, Prometheus, and Grafana Scale and load test Dapr applications on Kubernetes

Packt is searching for authors like you

If you're interested in becoming an author for Packt, please visit `authors.packtpub.com` and apply today. We have worked with thousands of developers and tech professionals, just like you, to help them share their insight with the global tech community. You can make a general application, apply for a specific hot topic that we are recruiting an author for, or submit your own idea.

Share Your Thoughts

Now you've finished *Embracing Microservices Design*, we'd love to hear your thoughts! Scan the QR code below to go straight to the Amazon review page for this book and share your feedback or leave a review on the site that you purchased it from.

`https://packt.link/r/1-801-81838-X`

Your review is important to us and the tech community and will help us make sure we're delivering excellent quality content.

Index

D